The Sulfur Problem: Cleaning up Industrial Feedstocks

RSC Clean Technology Monographs

Series Editor: J.H. Clark, *University of York, UK*

Advisory Panel: N.M. Edinberry (*Sandwich, UK*), J. Emsley (*London, UK*), S.M. Hassur (*Washington DC, USA*), D.R. Kelly (*Cardiff, UK*), T. Laird (*Mayfield, UK*), T. Papenfuhs (*Frankfurt, Germany*), B. Pearson (*Wigan, UK*), J. Winfield (*Glasgow, UK*)

The chemical process industries are under increasing pressure to develop environmentally friendly products and processes, with the key being a reduction in waste. This timely new series will introduce different clean technology concepts to academics and industrialists, presenting current research and addressing problem-solving issues.

Feedstock Recycling of Plastic Wastes
by J. Aguado, *Rey Juan Carlos University, Móstoles, Spain*; D.P. Serrano, *Complutense University of Madrid, Spain*

Applications of Hydrogen Peroxide and Derivatives
by C.W. Jones, *formerly of Solvay Interox R & D, Widnes, UK*

Clean Synthesis Using Porous Inorganic Solid Catalysts and Supported Reagents
by J.H. Clark and C.N. Rhodes, *Clean Technology Centre, Department of Chemistry, University of York, UK*

The Sulfur Problem: Cleaning up Industrial Feedstocks
by D. Stirling, *University of Glasgow, UK*

How to obtain future titles on publication

A standing order plan is available for this series. A standing order will bring delivery of each new volume upon publication. For further information please contact:

Sales and Customer Care
Royal Society of Chemistry
Thomas Graham House
Science Park
Milton Road
Cambridge
CB4 0WF
Telephone: +44(0) 1223 420066

RSC
CLEAN TECHNOLOGY
MONOGRAPHS

The Sulfur Problem: Cleaning up Industrial Feedstocks

Diane Stirling
Chemistry Department, University of Glasgow, UK

RS•C

ROYAL SOCIETY OF CHEMISTRY

ISBN 0-85404-541-4

A catalogue record for this book is available from the British Library

Published by The Royal Society of Chemistry,
Thomas Graham House, Science Park, Milton Road,
Cambridge CB4 0WF, UK

For further information see our web site at www.rsc.org

Typeset by Paston PrePress Ltd, Beccles, Suffolk
Printed and bound by MPG Books Ltd, Bodmin, Cornwall

Preface

Many industrially important feedstocks are contaminated with sulfur compounds, ultimately released into the atmosphere as sulfur oxides. The sulfur oxides are converted to acid rain, which damages the whole ecosystem, attacking vegetation and stonework, and acidifying lakes and rivers with detrimental effects on aquatic life. Current technologies for removing these sulfur contaminants at both percentage and ppm levels from industrial feedstocks are discussed. These are then linked to new materials which are currently being developed for sulfur clean-up. The synthesis, characterisation and testing of these materials are discussed in depth. Elements of solid state chemistry, the properties of small particles in solution (important in the synthesis step), the characterisation of materials and the kinetics of the sorption of gases in solids are included to place the clean-up of sulfur in industrial feedstocks on a sound theoretical footing.

Contents

Acknowledgements

I am greatly indebted to Dr G.W. Noble for reproducing the diagrams for this manuscript. I would also like to thank the following people for reading all or sections of the book: Professor J.M. Winfield, Dr K.C. Campbell and Dr B. Al-Duri.

CHAPTER 1

Introduction: the Sulfur Problem

1 Sources of Sulfur and Major Uses

Sulfur has been known since the beginning of history. It occurs uncombined in nature, and it is a major global pollutant when oxidised to sulfur dioxide. Sulfur compounds are used extensively in the modern industrialised world. Their main use is for the synthesis of sulfuric acid by the contact process.[1] In this, sulfur or sulfide minerals are converted to sulfur dioxide by heating them in air. The sulfur dioxide is then oxidised to sulfur trioxide in air over a supported potassium sulfate promoted vanadia catalyst at *ca.* 500 °C. The sulfur trioxide is absorbed in 98% sulfuric acid in ceramic packed towers and diluted to the desired concentration with water. Sulfur trioxide cannot be allowed to react directly with water, since this would result in the formation of a mist of sulfuric acid droplets that would pass right through the absorber and into the atmosphere. The contact process is very efficient, accounting for less than 2% of emissions into the atmosphere as sulfur dioxide. Sulfuric acid is mainly used in fertilisers. Other sulfur compounds are also used in a range of other industries including rubber vulcanising, leather processing and in the production of paper, cellulose, rayon, and many pharmaceuticals, fungicides and insecticides.

Hydrogen sulfide and organic sulfides are found in a variety of feedstocks. One of the major sources is the solid fossil fuel feedstock coal, which is derived from the partial degradation of plants. Coal can contain 0.1 to 6 wt% sulfur depending on its source. The chemical composition of coal is complex and non-stoichiometric but is typically comprised predominantly of carbon, hydrogen and oxygen with smaller amounts of nitrogen and sulfur. Bituminous coal, which contains 80% carbon, has a high calorific value but also a high sulfur content. Low sulfur coals such as anthracite are more desirable but supplies of these are diminishing as they have already been heavily mined. Approximately half of the sulfur in a typical coal will be found as pyrites (FeS_2). The remainder is found as organically bound sulfur, sulfate and H_2S.[2]

Another source of sulfur is crude oil. Crude is comprised predominantly of hydrocarbons with smaller amounts of sulfur, nitrogen and oxygen, and traces of heavy metals such as vanadium and nickel. The sulfur content of crude oil

Table 1.1 *Representative fractions from distillation of petroleum*
(Reproduced from ref. 3, p. 343, with permission from C.N. Satterfield)

Fraction	Component and/or boiling point range (°C)	Typical use
Gas	Up to C_4	Burned as fuel. Ethane may be thermally cracked to produce ethylene. Propane or a mixture of propane and butane may be sold as liquefied petroleum gas (LPG).
Straight-run gasoline	C_4–C_5	Blended into gasoline, isomerised or used as a chemical feedstock.
Virgin naphtha (light distillate)	C_5–150	Used as a feed to catalytic reformer or blended into gasoline.
Heavy naphtha (kerosene)	120–200 (Up to $\sim C_{15}$)	Jet fuel, kerosene.
Light gas oil	200–310 (Up to $\sim C_{20}$)	Used as a no. 2 distillate fuel oil, or blending stock for jet fuel and/or diesel fuel.
Gas oil (heavy distillate)	Up to ~ 350 ($\sim C_{25}$)	Used as a feed to catalytic cracker or sold as heavy fuel oil.
Atmospheric residual	$\sim 350 +$	Various uses. May be distilled under vacuum to produce vacuum gas oil, coked, or burned as fuel.
Vacuum residual	$\sim 560 +$ equivalent boiling point	Various. Some is hydrotreated catalytically.

depends on its origin. North African crudes contain *ca.* 0.2 wt% sulfur, midcontinental US crudes contain *ca.* 0.2–2.5 wt% sulfur and Venezuelan crudes 2–4 wt% sulfur.[3] The sulfur in crude oil is present as organic sulfur compounds, H_2S and small amounts of elemental sulfur. When processing crude oil it is separated into fractions by distillation. The fractions are defined by either their boiling point range or carbon chain length as appropriate; the fractions and their typical uses are detailed in Table 1.1. The atmospheric residual fraction is the residue left after distillation at 350 °C at 1 atmosphere. At higher temperatures, the crude oil is distilled under vacuum to form heavy gas oils, and the residue left from this is known as the vacuum residual. It has an equivalent boiling point of ~ 560 °C at 1 atmosphere pressure.

The sulfur content increases with increase in the overall molecular weight of the fraction. The main sulfur compounds are organic sulfides or disulfides, mercaptans and thiophenes in the low boiling fractions. The sulfur is found mainly as thiophene derivatives such as benzo- and dibenzothiophenes in the higher boiling fractions.

Other anthropogenic sources of sulfur include: (i) industrial gas streams which contain sulfur as carbonyl sulfide, carbon disulfide, low molecular weight mercaptans and thiophene; (ii) natural and refinery gases which contain sulfur as mercaptans, COS and thiophene; (iii) synthesis gas (CO + H_2) containing sulfur as COS and CS_2; and (iv) emissions from vehicle exhausts.[4]

Emissions from car exhausts have been much reduced in recent years by fitting three-way catalytic converters. The catalysts are comprised of platinum and rhodium dispersed on ceria-alumina mixed oxides coated on a monolith.[5] The monolith is a magnesia-alumina silicate that has been extruded into a series of parallel channels. The role of the catalyst is to effect the three-way conversion of small hydrocarbons, carbon monoxide and nitric oxide formed in the exhaust of the petrol engine into water, carbon dioxide and nitrogen. This will lower emissions of volatile organic carbon and nitrogen oxides which contribute to acid rain formation. However, sulfur may be emitted as H_2S and COS rather than SO_2 under these conditions, and H_2S is a known neurotoxin.[6] COS is very unreactive and is retained in the troposphere for a long time. The troposphere is the lower half of the earth's atmosphere. The upper layer of the atmosphere is known as the stratosphere and begins at around 15 km above the surface of the earth. Hydrocarbons and CO are converted to CO_2 and thus contribute to greenhouse gas emissions and thus global warming. It is clear that there are no easy solutions to control pollution levels in the environment.

There are also substantial natural reserves of sulfur, the most important of which are biogenic sources, sea spray and volcanoes. The biogenic sources originate from bacterial reduction of sediments to H_2S in the sea and release of dimethyl sulfide from sea organisms.[7] Most of the H_2S is redeployed in bacterial oxidation, so that dimethyl sulfide is the major biogenically produced sulfur compound in the oceans, giving rise to a concentration of 0.01 ppb sulfur in sea spray from this source.[8]

Volcanoes are the main natural source of sulfur dioxide. There are over 550 volcanoes in the world that are classed as active since they have erupted during historic time. When a volcanic eruption takes place, molten rock rises to the surface and it either flows in streams of glowing lava, or the molten material is violently ejected into the atmosphere together with large amounts of volcanic ash. Gases are also emitted during the eruption. The gases are mainly comprised of water vapour but are also accompanied by variable amounts of nitrogen, carbon dioxide and sulfur gases (mostly SO_2 and H_2S).[9] Small quantities of CO, H_2 and chlorine are also sometimes formed. Figure 1.1 shows a photograph of the Puu Oo vent from the Kilauea volcano in Hawaii and was taken in September 1983. The Kilauea volcano is the most active in the world and the photograph shows a stream of lava erupting from the vent of the volcano.[10]

The molten rock, which is also called the magma, only contains a few percent of these gases, but they can have a catastrophic effect on the environment. The eruption of Laki fissure in Iceland in 1783 is one of the earliest recorded volcanic eruptions.[11] Toxic gases, presumably SO_2, were emitted when the volcano erupted and they formed a haze which reduced the sunlight intensity right across Europe. Three quarters of the livestock in Iceland died, and the haze

Figure 1.1 *Eruption of the Puu Oo cone of the Kilauea volcano.*
(Photograph by J.D. Griggs, September 1983. Used with permission from
U.S. Geological Survey, Hawaiian Volcano Observatory)

caused the air temperature to drop so that the crops failed. This was then
followed by the coldest winter in 225 years. There was insufficient food to
support the population and *ca.* 24% of the Icelanders died of starvation. A more
recent example is the eruption of Mount Pinatubo in the Philippines in June
1991. Approximately 20 million tonnes of SO_2 along with 3 to 5 km^3 of
particulate matter (mostly volcanic ash) were ejected into the atmosphere.[9]
One meteorological report showed that 1992 temperatures were at a ten year
low in the northern hemisphere and a fifteen year low in the southern hemisphere following the eruption. Volcanic eruptions can also create considerable
long term social disruption to nearby communities, as demonstrated by the
volcanic eruption of the Soufriere hills volcano on Montserrat which lasted
from July 1995 to March 1998. The entire population on the southern half of the
island had to be relocated.

   Although the effects of volcanic activity are devastating, natural sources, of
which volcanoes are the biggest component, account for less than 30% of sulfur
emissions, so that man is by far the biggest contributor to atmospheric sulfur
levels. The major anthropogenic source of SO_2 is from coal fired power stations.
This could be avoided by completely removing SO_2 from effluent gases, but this
is not economically or technologically achievable. However, many techniques
exist for substantially reducing SO_2 emissions, and they will be discussed in
Chapter 4 of this monograph. Other sources of SO_2 emissions include oil-
refinery operations, oil-fired energy generation, copper smelting and sulfuric
acid manufacture.[1]

The large scale industrial use of sulfur-containing gas, oil and solid fuel feedstocks makes it essential to clean up both feedstocks and refinery effluent in order to minimise production costs and prevent extensive pollution of the environment. Thus, sulfur contaminants emitted into the atmosphere ultimately form acid rain (Section 3), and sulfur contaminants found in industrial feedstocks cause plant pipeline corrosion at concentrations in excess of 3 ppm. Sulfur contaminants can also poison catalysts used in the processing of feedstocks. A nickel on alumina catalyst, for example, is used in steam reforming in which steam reacts with methane at 800 °C at 35 bar to form synthesis gas (carbon monoxide and hydrogen). The catalyst is severely poisoned by sulfur compounds which behave as Lewis-type bases donating electrons into the unfilled d orbitals of the metal.[12]

2 Environmental Sulfur Levels

Sulfur dioxide and sulfate concentrations have been determined across the UK in recent years. They show that SO_2 emissions have decreased from 240 000 metric tonnes (measured as S) in 1987 to 140 000 metric tonnes in 1992–1994. Sulfur as sulfate has decreased from 230 000 metric tonnes to 200 000 metric tonnes in the same period.[13] The average concentration of SO_2 in the atmosphere in the UK today is *ca.* 33 μg m^{-3}, most of which comes from anthropogenic sources; SO_2 emissions in urban areas are considerably higher than those in rural areas. This figure is considerably smaller than it was in the past; the average urban SO_2 concentrations were 188 μg m^{-3} in 1958, 144 μg m^{-3} in 1970 and 73 μg m^{-3} in 1977 in the UK.[8] This reduction in emissions is partly accounted for by the enforcement of the Clean Air Acts of 1956 and 1968 and conversion of housing to smokeless fuels, natural gas and electricity. Although emissions from low household chimneys have decreased, emissions from the taller chimneys associated with power stations have considerably increased. Taller chimneys will cause the sulfur to be carried away from the immediate vicinity so that localised sulfur levels will be reduced, but it will then be deposited elsewhere. In the absence of a source of moisture such as clouds, sulfur dioxide can travel hundreds of kilometres to eventually be precipitated as acid rain when it comes into contact with moist air. In Sweden, for example, where the effects of acid rain on lakes and streams has been quite severe, only one tenth of the pollution originates from atmospheric SO_2 emissions in Sweden; a further tenth can be attributed to UK sulfur emissions and the remainder is from industrial regions in northern Europe.[1] It has to be said, though, that now we are aware of this problem, legislation has been introduced to reduce/eliminate acid rain substantially. A series of environmental action programmes has been developed, which included remedial measures to reduce SO_2 emissions. New measures such as encouraging the use of 'cleaner' energy resources, recycling of waste and conservation of natural resources have also been introduced. One interesting effect of the reduction in sulfur emissions has been that soils in some areas are now so sulfur deficient that sulfur is having to be added to the fertilisers used on this ground!

Species other than SO_2 found in the atmosphere include H_2S, dimethyl sulfide $[(CH_3)_2S]$, dimethyl disulfide $[(CH_3)_2S_2]$, carbonyl sulfide (COS) and carbon disulfide (CS_2). H_2S, $(CH_3)_2S$ and $(CH_3)_2S_2$ are rapidly oxidised to SO_2 and thus have lifetimes of only a few days in the atmosphere. Carbonyl sulfide and carbon disulfide are much longer lived species and are found in the troposphere.[14] Carbonyl sulfide in particular is a problem since it is present in the troposphere at concentrations of *ca.* 500 μg m^{-3}. Concentrations of carbon disulfide are less than one tenth those of the carbonyl sulfide. H_2S, COS and CS_2 react with hydroxide radicals to form SH radicals in the atmosphere.[14]

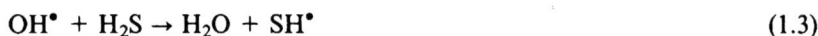

$$OH^\bullet + CS_2 \rightarrow COS + SH^\bullet \qquad (1.1)$$

$$OH^\bullet + COS \rightarrow CO_2 + SH^\bullet \qquad (1.2)$$

$$OH^\bullet + H_2S \rightarrow H_2O + SH^\bullet \qquad (1.3)$$

The SH radicals are then further oxidised to SO_2.

The hydroxide free radicals are formed in the atmosphere by photolysis of ozone. This involves the splitting of an ozone molecule by the absorption of solar radiation, followed by reaction with water to give the hydroxide free radicals.[8]

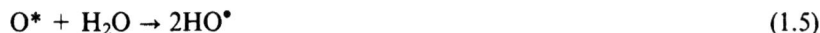

$$O_3 + h\nu \rightarrow O_2 + O^* \qquad (1.4)$$

$$O^* + H_2O \rightarrow 2HO^\bullet \qquad (1.5)$$

It is important not to consider sulfur emissions in isolation, as they are accompanied by other pollutant gases resulting from domestic and industrial stack emissions and emissions from motor fuel combustion. Other main pollutants include smoke and particulates, nitrogen oxides, CO, CO_2, HCl, hydrocarbons and heavy metals. Hydrocarbons, heavy metals, smoke and particulates are a health hazard, particularly to asthmatics, nitrogen oxides contribute (along with SO_2) to acid rain formation (see Section 3), and CO_2 contributes to global warming. HCl and chlorine compounds, used in aerosol spray cans, refrigerants and solvents, have increased the amount of chlorine in the atmosphere, and this breaks down to chlorine radicals which damage the ozone layer. Ozone breakdown occurs in the stratosphere. Chlorofluorocarbons (CFCs) were used for many years as inert, non-toxic, non-flammable compounds in refrigeration, for example. Unfortunately, they decompose photochemically in the stratosphere to chlorine which again catalyses the decomposition of ozone. CFCs were phased out completely from 1st January 1996. Since then, more 'environmentally friendly' non-flammable, non-toxic replacements have been sought for use as refrigerants. These include hydrofluorocarbons (HFCs) and hydrochlorofluorocarbons (HCFCs). These materials break down in the troposphere where there is very little ozone. However, HCFCs still contain chlorine which is released into the troposphere when the HCFCs break down and some of this chlorine may find its way into the stratosphere and therefore still contribute to ozone breakdown. The best alternative CFC replacements are therefore the HFCs and indeed, the HCFCs have now been largely phased out.[15] The main effect of the depletion of the ozone layer is to increase the amount of ultraviolet radiation reaching the earth's surface leading to an increased risk of eye cataracts and skin cancer in the human population.

Sulfur dioxide is a respiratory irritant at concentrations > 1 ppm, especially when it is combined with soot. It can cause pleurisy, bronchitis and emphysema in susceptible individuals.[14] Its effects were particularly marked before legislation was introduced to stop coal being used as a fuel in cities. Most British coal is bituminous and has a high tar and hydrocarbon content. This causes it to smoke considerably when it is burned. Smoke in a humid environment, such as that commonly found in British winters, acts as nucleation centres for fog formation, and this combination of smoke and fog is known as smog. One of the worst incidents from smog pollution occurred in London in 1952 when four thousand people died. A cold black sulfurous smog was trapped over the city by a blanket of warm air for almost a week. This irritated the bronchial tubes of individuals inhaling the smog so that they flooded with mucus and the people choked to death. It was thought that the smoke and sulfur dioxide caused the deaths, but it is now thought that the effects of the sulfurous smog were accentuated by the formation of highly acidic particles.[14]

3 Acid Rain

The sulfur dioxide that is released into the atmosphere generates acid rain. Acid rain is extremely detrimental to the environment, causing lakes and soils to become acidified and resulting in the death of fish and the poisoning of trees. Acid rain also damages buildings, particularly limestone buildings, by accelerating the rate of decay of stone and increases the rate of corrosion of metal structures such as bridges. Acid rain was first identified by Robert Angus Smith in England in 1872, but the detrimental effects of acid rain were not identified until 1961. Since then steps have been taken to limit acid rain formation by controlling gas emissions into the atmosphere. Although sulfur dioxide is the main precursor to acid rain formation, nitrogen oxides are also involved in acid rain formation. Traces of HCl are also found in acid rain, and they are thought to originate from the reaction of sulfuric acid with atmospheric sodium chloride originating from evaporation of seawater.[14]

Sulfur dioxide in the atmosphere can be oxidised by a variety of routes, *e.g.* gas–gas reactions such as the interaction of hydroxide free radicals with SO_2 to generate $HSO_3^{\bullet}$:[8]

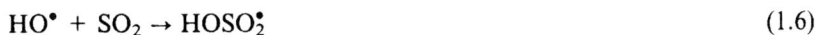

$$HO^{\bullet} + SO_2 \rightarrow HOSO_2^{\bullet} \tag{1.6}$$

$HOSO_2^{\bullet}$ is very reactive and is eventually further oxidised to sulfate either in water droplets or on nitrogen-containing molecules such as NO.[2] Reactions also take place in solution and may involve a catalyst such as manganese,[8] *e.g.*:

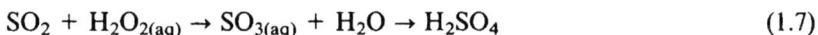

$$SO_2 + H_2O_{2(aq)} \rightarrow SO_{3(aq)} + H_2O \rightarrow H_2SO_4 \tag{1.7}$$

The H_2O_2 is formed from the combination of two $HO^{\bullet}$ radicals. Most of the sulfate that is formed is then washed to earth in rain or snow, but some sulfur compounds (a mixture of SO_2 and sulfate) are adsorbed on moist surfaces such as soil and vegetation. In the absence of moisture, the sulfur oxides and sulfates can travel up to 4000 km from their source, but most are precipitated out from rain/snow within 200 km of their source.[8]

Ordinary rainwater is slightly acidic due to dissolved atmospheric CO_2; it has a pH of 5.6. Acid rain is defined as rainwater with pH < 5. Lakes become acidified by drainage of rainwater from the soil. The sulfate ion found in acid rain is much more effective in transferring acidity from soils (by binding to a hydrogen ion) to surface water than the carbonic and organic acids found naturally in soils. Soils also contain large amounts of insoluble aluminium, and acid rain solubilises the aluminium so that it is washed into streams and lakes. Aluminium and other toxic metals released from the soil in these acidic conditions prevent fishes' gills from functioning properly and the fish die. The acidified soils also have a detrimental effect on trees. Direct damage to the leaves occurs by the absorption of SO_2 and acid rain. This affects transpiration and photosynthesis and can cause leaf loss. More extensive acid rain damage first of all causes soluble metallic nutrients such as calcium and magnesium to be leached out from the soil and at a later stage aluminium is extracted from the soil and poisons the trees.[8]

Acid rain also has a detrimental effect on buildings. It accelerates the rate of decay of calcite ($CaCO_3$) which is found in limestone,[16] *i.e.*:

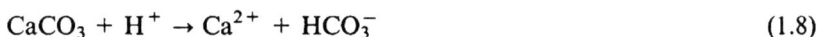

$$CaCO_3 + H^+ \rightarrow Ca^{2+} + HCO_3^- \tag{1.8}$$

This reaction has a large positive equilibrium constant, driving the reaction in the forward direction so that soluble Ca^{2+} salts are washed out of the stonework causing pitting. SO_2 adsorbed on the surface of limestone in dry climates can also cause damage to buildings. The SO_2 is oxidised and reacts with the $CaCO_3$ to form gypsum:

$$CaCO_3 + SO_2 + \tfrac{1}{2}O_2 + 2H_2O \rightarrow CaSO_4 \cdot 2H_2O + CO_2 \tag{1.9}$$

Gypsum can hydrate and form a crust on the surface of the stonework.

Urban stone decay is fastest in humid atmospheres that contain ozone, nitrogen oxides and SO_2. The oxidation of SO_2 on stone surfaces is catalysed by NO_2.[16] The use of silane polymer coatings to protect limestone from acid rain is currently being investigated. The surface is first cleaned by sandblasting or by using strong acids or alkalis. The protective coating can then be applied. A major problem with this treatment is that these coatings are often impermeable so that salts emitted from the stonework will build up on the surface rather than be washed away, and they will eventually break through the polymer coating. The stonework is then once again exposed to pollutants.

4 Conclusions

Clearly, sulfur-containing pollutants are a major threat to the environment, and anthropogenic sources arising from fossil fuels, industrial gas streams and vehicle emissions now vastly outweigh sulfur pollutants from natural sources, *i.e.* biogenic activity, sea sprays and volcanoes. Sulfur emissions are generally measured in terms of sulfur oxides, since H_2S and the organic sulfides (dimethyl sulfide and dimethyl disulfide) are rapidly oxidised to SO_2. Although measures are now underway to control SO_2 emissions, the large increase in instances of childhood asthma and the continuing detrimental effects of acid rain on the environment indicate that the effects of sulfur on the environment will continue to be a problem for years to come. Furthermore, with the presence of concentrations of *ca.* 500 μg m^{-3} of long-lived COS species in the troposphere and the interactions of sulfur-containing species with other gases in the atmosphere, we may well be storing up as yet unforseen problems for ourselves in the future.

5 References

1 N.N. Greenwood and A. Earnshaw, 'Chemistry of the Elements', Pergamon Press, Oxford, 1990.
2 S.E. Manahan, 'Environmental Chemistry', 5th edn, Lewis Publishers, Michigan, 1991.
3 C.N. Satterfield, 'Heterogeneous Catalysis in Industrial Practice', 2nd edn, McGraw Hill, New York, 1991.
4 A. Kohl and F. Riesenfeld, 'Gas Purification', 4th edn, Gulf Publishing Company, Houston, 1985.
5 J.M. Thomas and W.J. Thomas, 'Principles and Practice of Heterogeneous Catalysis', VCH, Cambridge, 1997.
6 S.F. Watts and C.N. Roberts, *Atmospheric Environment*, 1999, 33, 169.
7 P. Kelly, *Chem. Brit.*, 1997, 33, 25.
8 P. O'Neill, 'Environmental Chemistry', 2nd edn, Chapman and Hall, London, 1993.
9 J.S. Monroe and R. Wicander, 'Physical Geology. Exploring the Earth', 2nd edn, West Publishing Company, Minneapolis, 1995.
10 J.D. Griggs, photograph of Puu Oo cone, Kileau Volcano, Hawaii (US Geological Survey, Hawaiian Volcano Observatory).
11 H.R. Bárdason, 'Ice and Fire', 4th English Edn., H.R. Bárdason, Reykjavík, 1991.
12 C.H. Bartholomew, P.K. Agrawal and J.R. Katzer, *Adv. Catal.*, 1982, 31, 135.
13 'Acid deposition in the United Kingdom, 1992–1994', 4th report of the review group on acid rain, AEA Technology plc, 1997, p. 129.
14 R.P. Wayne, 'Chemistry of Atmospheres, an Introduction to the Chemistry of the Atmospheres of Earth, the Planets and their Satellites', Clarendon Press, Oxford, 1985.
15 G. Webb and J.M. Winfield, 'CFC alternatives and new catalytic methods of synthesis', Chapter 8 in 'Chemistry of Waste Minimisation', ed. J.H. Clark, 1995, Chapman and Hall, Cambridge, 1995, p. 222.
16 G. Allen and J. Beavis, *Chem. Brit.*, 1996, September, 24.

CHAPTER 2

Catalytic Hydrodesulfurisation

1 The Process

Although some of the organic sulfur compounds found in oil and other feedstocks can be removed by the absorption, adsorption and oxidation processes that are used for H_2S removal (Chapters 3 and 4), organic sulfur compounds are generally much less reactive than H_2S. A high temperature hydrodesulfurisation reaction is therefore needed to convert the organosulfides to H_2S.

Hydrodesulfurisation (HDS) is the removal of sulfur by a reduction treatment. Sulfur present as thiols, sulfides, disulfides and thiophenes in oil feedstocks undergoes hydrogenolysis to generate H_2S and a hydrocarbon, *e.g.* for methyl mercaptan:

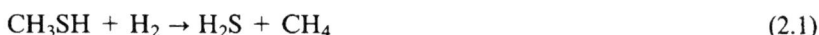

$$CH_3SH + H_2 \rightarrow H_2S + CH_4 \tag{2.1}$$

and for thiophene:

 $+ 3H_2 \rightarrow H_2S + C_4H_8$ $\tag{2.2}$

For given R and R' groups, the reactivity decreases in the sequence:

$$RSH > RSSR' > RSR' > \text{thiophenes}$$

where R is an aliphatic or aromatic group. High molecular weight organosulfides are less reactive than low molecular weight organosulfides, activity decreasing with increasing molecular weight.[1]

Hydrodesulfurisation is one of a number of hydrotreating processes used in treating feedstocks. All the processes involve reaction with hydrogen. Other processes include removal of nitrogen by hydrodenitrogenation (HDN), removal of oxygen (as water) from oxygen containing compounds by hydrodeoxygenation (HDO), removal of heavy metals by hydrodemetallisation (HDM), removal of chlorine by hydrodechlorination and hydrogenation of unwanted unsaturated compounds.

Hydrodesulfurisation is carried out over a presulfided, alumina supported, cobalt or nickel molybdate catalyst at *ca.* 350 °C and at 30 to 50 bar.[1] The cobalt molybdate catalyst is generally used in preference to the nickel molybdate catalyst since it is more effective at hydrogenolysis.[2] Nickel molybdate is the preferred option for HDN since it is a better hydrogenation catalyst; more hydrogen is consumed in HDN since the reaction is run at higher temperatures and is generally applied to the heavier fractions of crude oil which contain substantial amounts of unsaturated products, since they contain the most nitrogen.[1]

The hydrogenolysis of organosulfides is exothermic. However, the concentration of the organosulfides is sufficiently low in most feedstocks for any temperature rise in the catalyst bed to be generally negligible.[2] As stated above, the catalyst is sulfided prior to use. Sulfiding can be carried out at *ca.* 300 °C using a feed gas typically containing *ca.* 1% of an easily decomposed sulfur compound such as dimethyl sulfide and *ca.* 5% hydrogen.[2] During this activation step the MoO_3 is converted to MoS_2 and the cobalt is partially sulfided. It is important that the dimethyl sulfide is introduced to the catalyst at the same time as the hydrogen since, if the catalyst is prereduced to MoO_2, it is less readily sulfided and large amounts of oxygen are retained in the catalyst.[1] The catalyst formed under these conditions is also less active and selective in HDS. Since the HDS reactions are exothermic they are favoured by low temperatures and high pressures. The actual operating conditions are a compromise between the thermodynamically favoured temperatures and pressures and conditions compatible with plant operations. A temperature of *ca.* 350 °C is generally used because the catalysts are inactive below 280 °C and hydrocarbon cracking reactions occur at temperatures greater than 400 °C. The reaction is carried out at 30–50 bar, depending on the operating pressure of the plant, in a feed gas which contains 2–5% H_2 in a natural gas feedstock and 25% H_2 if using a feedstock containing higher hydrocarbons such as naphtha. Naphtha is a general term used to describe a mixture of hydrocarbons formed from the fractional distillation of oil feedstocks. The light distillate fraction (see Table 1 in Chapter 1) for C_5 hydrocarbons, for example, is known as virgin naphtha and the kerosene fraction is referred to as heavy naphtha. When using naphtha feedstocks, the excess hydrogen is used to suppress cracking reactions.[2]

2 The Catalyst

The $Co\text{-}Mo/Al_2O_3$ hydrodesulfurisation catalyst is prepared by impregnation of a high surface area alumina support with aqueous solutions of $Co(NO_3)_2{\cdot}6H_2O$ and $(NH_4)_6Mo_7O_{24}$.[3] The catalyst typically contains 1–4 wt% Co and 8–16 wt% Mo. The catalyst precursor is then dried and calcined to give the supported mixed Co/Mo oxide which is then activated by sulfiding in

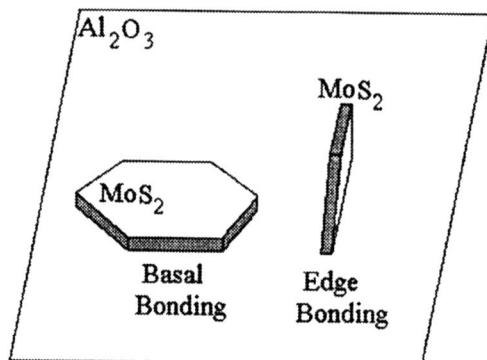

Figure 2.1 *Orientation of small MoS₂ crystallites on the surface of Al₂O₃.*
(Reprinted from ref. 3, figure 1, page 5, courtesy of Marcel Dekker Inc.,
New York)

hydrogen and a sulfur-containing feed as described in Section 1. The precise
composition of the active catalyst is governed by the concentration and the
order of impregnation of the metal salts, the calcination temperature and the
degree to which the catalyst is sulfided.[4] A calcination temperature of *ca.* 500 °C
is used for optimum performance. At higher temperatures more molybdenum
and cobalt aluminates are formed and molybdenum oxide is volatile at
temperatures above 700 °C.[5] The MoO_3 is almost completely sulfided during
activation to form MoS_2. MoS_2 has a layered structure with weak van der Waals
type interactions between the sulfur atoms in contiguous layers.[1] The layers can
be bonded to the alumina surface either by their basal plane or on their edges
(Figure 2.1).[3]

Surface science studies using a MoS_2 single crystal showed that this had
negligible activity for the hydrodesulfurisation of thiophene.[5] The surface area
of the basal plane of the crystal was much greater than that of the edge. When
the crystal was bombarded with Ar^+ (sputtered) so that molybdenum ions were
exposed, there was a marked increase in the hydrodesulfurisation activity.[6] This
indicates that the active sites for catalysis are formed at the corners or edges of
layers rather than the basal planes. The main role of the alumina is to stabilise
the MoS_2 layers.

The cobalt in the Co-Mo/Al_2O_3 catalyst prior to sulfiding can occupy
tetrahedral or octahedral sites in the alumina and/or form Co_3O_4 crystallites
on the surface.[3] The exact location will depend on the cobalt loading and the
treatment of the impregnated precursor. On sulfiding, the cobalt is found in
three environments (Figure 2.2).

Co_3O_4 forms Co_9S_8; cobalt ions in tetrahedral sites in the alumina remain
there after sulfidation and cobalt ions that were located in octahedral sites in
the oxides are adsorbed on the edges of the MoS_2 crystallites after sulfiding.[3,4]
The cobalt found on the edges of the MoS_2 layers is known as the Co-Mo-S
phase,[3] and this is thought to be the principal active site in the hydrodesulfurisation catalyst. The cobalt ions in the tetrahedral sites of the alumina are few

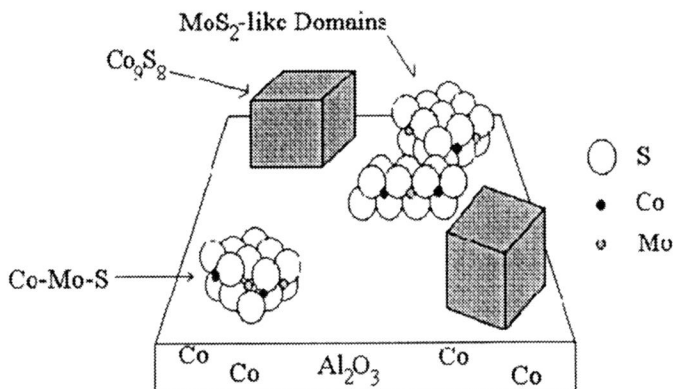

Figure 2.2 *Schematic representation of the different phases present in a typical alumina-supported catalyst.*
(Reprinted from ref. 4, figure 3, page 401, courtesy of Marcel Dekker Inc., New York)

in number[4] and are thought to be inactive.[1] The proportion of the phase Co_9S_8 increases with increasing cobalt loading.[4] The catalytic activity for hydrodesulfurisation of the sulfided $Co-Mo/Al_2O_3$ catalyst is much greater than that of Mo/Al_2O_3, and there is some controversy concerning the precise role of the cobalt in the active Co-Mo-S phase. It has been suggested that the cobalt acts as a promoter by improving hydrogen adsorption,[1] or by increasing the number of active molybdenum sites at the surface.[3] It has also been proposed that the cobalt may replace the molybdenum as the catalytically active site.[3]

The Co-Mo-S phase is found in two forms. Type I is formed after low temperature activation in H_2S. In this structure the Co-Mo-S is bonded to the support by oxygen bonds. The type II Co-Mo-S phase is formed by high temperature sulfiding when all these bonds are either broken or sulfided. The precise temperature at which the change occurs decreases with increasing Co/Mo ratio.[5] The type II structures are more active for HDS whereas the type I structures are more active for hydrogenation reactions.

3 Mechanism of Hydrodesulfurisation

The model compound thiophene is selected for a discussion of the mechanism since it is typical of sulfur-containing compounds found in feedstocks. Although thiophene has been widely studied as a model sulfur compound in hydrodesulfurisation, there is still considerable controversy about the mechanism of desulfurisation. Three mechanistic routes have been identified.[5,7]

The first route entails the hydrogenation of the thiophene ring. The thiophene molecule is regarded as a pseudo aromatic molecule due to delocalisation of electrons through its π system and this gives it an inherent stability. Partial or complete hydrogenation of the ring results in a loss of this aromatic character and stability, *i.e.*:

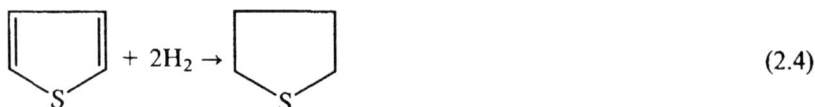

$$\text{(structure)} + H_2 \rightarrow \text{(structure)} \tag{2.3}$$

$$\text{(structure)} + 2H_2 \rightarrow \text{(structure)} \tag{2.4}$$

The second route involves the cleavage of a carbon–sulfur bond and addition of hydrogen to give a diene thiol intermediate.

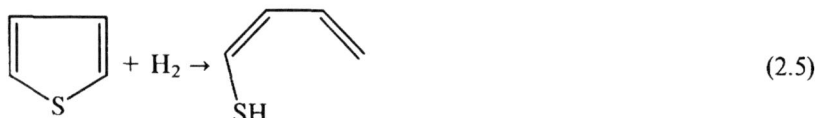

$$\text{(structure)} + H_2 \rightarrow \underset{\text{SH}}{\text{(structure)}} \tag{2.5}$$

An elimination reaction then takes place with the diene thiol intermediate to form butadiene and H_2S.[7] Butadiene can then be further hydrogenated to butenes and butane.

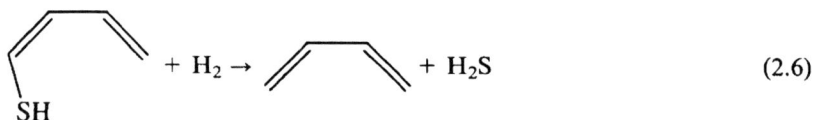

$$\underset{\text{SH}}{\text{(structure)}} + H_2 \rightarrow \text{(structure)} + H_2S \tag{2.6}$$

The cleavage and elimination reactions described above can either take place in two stages as shown above, or by a concerted reaction in which the intermediates are retained on the surface of the catalyst and only butenes or butane are observed as products.[5]

   The third route for the desulfurisation of thiophene involves dehydrosulfurisation of thiophene to form butadiyne. This is an elimination and cleavage reaction which takes place without involvement of the hydrogen. It is thought that this reaction is unlikely to occur for thiophene due to the stability of the thiophene ring. This reaction typically occurs in alkyl sulfides where the carbon–sulfur bonding is less stable.

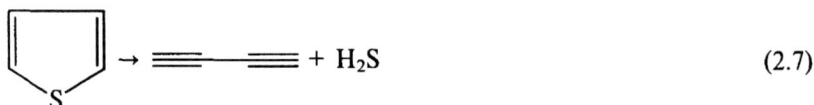

$$\text{(structure)} \rightarrow \text{(structure)} + H_2S \tag{2.7}$$

In practice, it is unlikely also that cleavage of a C–S bond would take place directly from thiophene as it is too stable. The most likely mechanism is therefore that the thiophene will be partially hydrogenated, as shown in Equation 2.3; this would destabilise the ring so that the molecule is more susceptible to C–S bond cleavage and elimination of H_2S as described in Equations 2.5 and 2.6.

The adsorption sites for the thiophene are thought to be anion vacancies present in the Co-Mo-S phase.[4] Theoretical calculations indicate that the thiophene is adsorbed perpendicularly to the surface via the sulfur. Hydrogen is thought to adsorb on a different active site, and there is considerable controversy as to whether the hydrogen is present as a molecular or dissociatively adsorbed species. The rate-determining or slow step in hydrodesulfurisation is the reaction between adsorbed thiophene and adsorbed hydrogen.[8] There is also evidence for the participation of a further adsorption site in which intermediate butenes desorbed from the anion vacancy sites are adsorbed and further hydrogenated to butane.[9]

4 References

1 C.N. Satterfield, 'Heterogeneous Catalysis in Industrial Practice', 2nd edn, McGraw Hill, New York, 1991.
2 P.J.H. Carnell, Chapter 4 in 'Catalyst Handbook', ed. M.V. Twigg, Wolfe Publishing Limited, Frome, England, 1989.
3 R. Prins, V.H.J. de Beer and G.A. Somorjai, *Catal. Rev. Sci. Eng.*, 1989, **31**, 1.
4 H. Topsoe and B.S. Clausen, *Catal. Rev. Sci. Eng.*, 1984, **26**, 395.
5 H. Topsoe, B.S. Clausen and F.E. Massoth, 'Catalysis Science and Technology, Volume 11', ed. J.R. Anderson and M. Boudart, Springer Publishers, Berlin, 1996, and references therein.
6 M. Salmeron, G.A. Somorjai, A. Wold, R.R. Chianelli and K.S. Liang, *Chem. Phys. Lett.*, 1982, **90**, 105.
7 M. Zdrazil, *Appl. Catal.*, 1982, **4**, 107.
8 M.L. Vrinat, *Appl. Catal.*, 1983, **6**, 137.
9 C.N. Satterfield and G.W. Roberts, *AIChE J.*, 1968, **14**, 159.

CHAPTER 3

Adsorption and Absorption of H₂S

1 Introduction

The hydrogen sulfide that is removed by adsorption and/or absorption processes originates from a number of anthropogenic sources. These include sewerage and municipal waste gases, and refinery processing gas streams. The processing gas streams include those arising from hydrodesulfurisation of organosulfides and other hydrorefining processes (see Chapter 2), and other refinery operations such as catalytic cracking and hydrocracking.

Catalytic cracking involves the breakdown of heavy oils into lighter fractions for processing into petrol and fine chemicals using a zeolite catalyst at a pressure of *ca.* 0.2–0.3 MPa and a temperature of *ca.* 520–540 °C.[1] Hydrocracking combines hydrogenation and catalytic cracking. It is used for breaking down heavy gas and vacuum oils which contain large amounts of polycyclic aromatic hydrocarbons into lighter fractions such as jet fuel. The hydrogen minimises coking and thus deactivation by hydrogenating coke precursors which build up on the surface of the catalyst. A zeolite supported Ni-Mo catalyst can be used for hydrocracking, and the operating conditions are usually *ca.* 200–400 °C at a hydrogen pressure of 1–10 MPa, according to the feedstock used.[1] Hydrogen sulfide is formed after cracking and hydrocracking by the breakdown of sulfur-containing molecules present in the oil feedstocks at the operating temperatures and pressures of the cracking plants.

The problem of removing hydrogen sulfide from the wide variety of feed-stocks discussed above is complicated by the fact that hydrogen sulfide is not found in isolation. Thus, in sewerage and municipal waste gases it is accompanied by methane, whereas in refinery gases hydrogen sulfide is found together with methane, hydrogen and higher hydrocarbons, and traces of nitrogen-, oxygen- and metal-containing species which have not been completely removed during processing. Nevertheless, the reduction in concentration of hydrogen sulfide from percentage to ppm levels in these process streams is readily achieved. Routes to this include: absorption in liquids such as alkanolamine,

16

ammonia solution and alkaline salt solutions; and oxidation of H$_2$S using iron(III) oxide, activated carbon or a Claus process. These processes are described in this chapter. The challenging problem is to remove the last traces of H$_2$S from the process stream. This is important since, as stated in Chapter 1, a few ppm of sulfur can corrode pipelines, poison catalysts used in fuel processing and can ultimately be emitted into the atmosphere as acid rain. A number of materials have been identified that are effective at completely removing H$_2$S from feed streams at high temperatures, but the ultimate goal is to develop materials that can remove H$_2$S at room temperature. This has the advantage that the materials can be readily transported to the process stream, even if it is in an inaccessible location such as an oil platform, and plant operating costs are kept to a minimum. Materials developed for either high or low temperature removal of H$_2$S are described in Section 3. The high surface area materials used to remove gases such as H$_2$S from feedstocks are referred to as 'sorbents' since adsorption of the gas onto the surface of the solid is followed by absorption of the sulfur into the bulk of the solid. Thus, the removal of H$_2$S is generally a two-stage process, the sulfur level being reduced to 2–3 ppm by an absorption or oxidation step in the first stage, and then in the second stage the sulfur content of the feedstock is reduced effectively to zero by sorption using a high surface area active solid sorbent. Many of these sorbents are also effective at higher concentrations of H$_2$S. Traces of organic sulfur compounds are left in feed-stocks after processing and these compounds are much less chemically reactive than H$_2$S and are therefore only partially eliminated by some of the adsorption/absorption processes described below.

2   Absorption into a Liquid

When H$_2$S is absorbed in a liquid, it can either dissolve or react chemically. Liquids used for absorption include alkanolamines, aqueous ammonia, alkaline salt solutions and sodium and potassium carbonate solutions.[2]

Absorption in Alkanolamine

Monoethanolamine, HOCH$_2$CH$_2$NH$_2$, is one of the most widely used alkano-lamines for H$_2$S removal. The function of the hydroxyl group is to decrease the vapour pressure and increase the water solubility of the alkanolamine, and the amine group makes an aqueous solution of the compound basic so that it can neutralise acidic gases such as H$_2$S. Monoethanolamine reacts with H$_2$S to form the amine sulfide and hydrosulfide,[1] *i.e.*:

$$2HOCH_2CH_2NH_2 + H_2S \rightleftharpoons (HOCH_2CH_2NH_3)_2S \qquad (3.1)$$

$$(HOCH_2CH_2NH_3)_2S + H_2S \rightleftharpoons 2HOCH_2CH_2NH_3HS \qquad (3.2)$$

A schematic diagram of the plant used for H$_2$S absorption using an aqueous

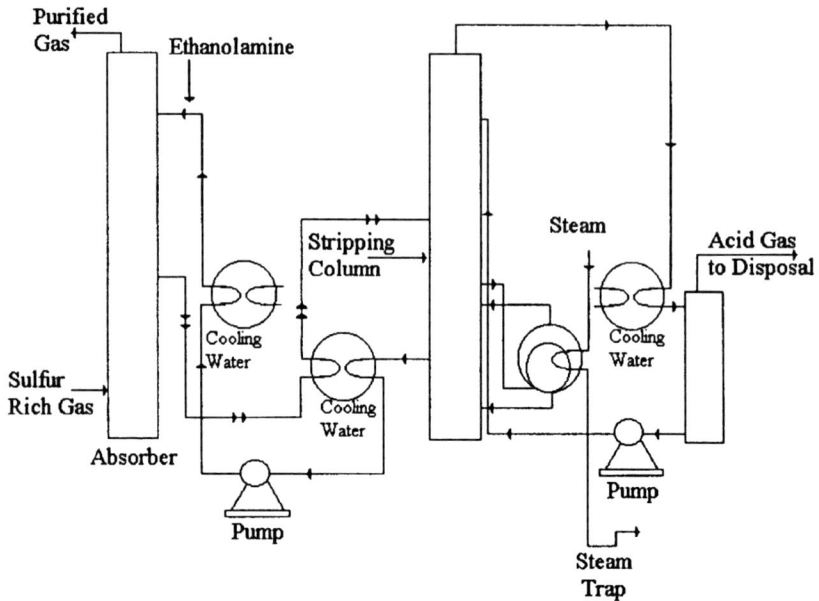

Figure 3.1   *Schematic diagram for alkanolamine acid gas removal processes.*
(Adapted from ref. 1. Used by permission. Adapted from *Gas Purification*
© 1985, Gulf Publishing Company, Houston, Texas, 800-231-6275. All
rights reserved)

solution of monoethanolamine is shown in Figure 3.1. The sulfur-containing
feedstock stream is passed up through an absorber bed. The monoethanolamine solution flows countercurrently to the gas stream. Absorption columns
contain either packing or trays in order to increase the interfacial contact and
hence mass transfer between the wash solution and the gas stream. Packing
can be either random or ordered, and one of the commonest types of packing
material is the Raschig ring (Figure 3.2); this is widely used in monoethanolamine absorption towers. The packed column is usually cylindrical, and it
contains a support plate for the packing material and a multi-exit distributor
for dispensing the absorbent liquid onto the top of the packing.[3] A schematic
diagram of a packed absorption tower is shown in Figure 3.3. Alternatively,
the packed column can contain trays and one of the most commonly used

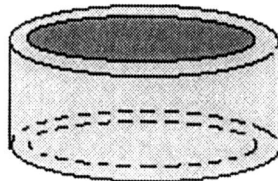

Figure 3.2   *Schematic diagram of a Raschig ring.*

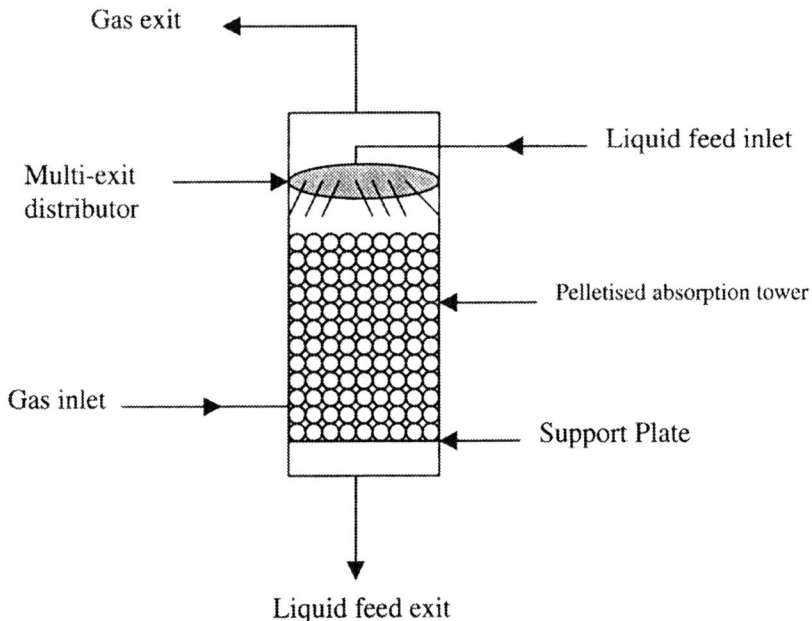

Figure 3.3   *Schematic diagram of a packed absorption column.*

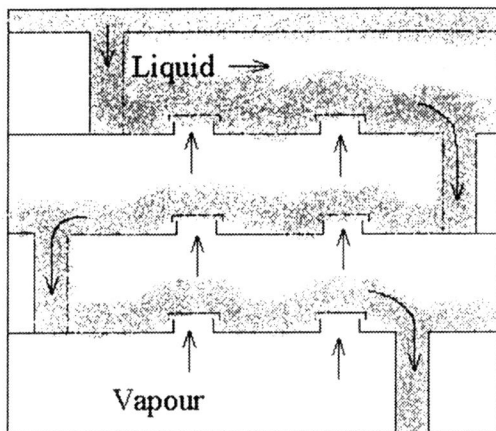

Figure 3.4   *Schematic diagram of a bubble cap absorber column.*

types is the bubble cap. A schematic diagram of this is given in Figure 3.4. In the bubble cap absorber column, the absorber liquid cascades down the column in a series of layers whereas the feedstock vapour bubbles up through each of the layers. Mass transfer then takes place at the surfaces of the bubbles.

The monoethanolamine solution that collects at the bottom of the absorber bed is sulfur rich whereas the cleaned feedstock stream passing out of the top of the absorber contains only minute traces of sulfur. The used monoethanolamine solution passes next, via a heat exchanger, to the top of a stripping column (Figure 3.1). The role of the stripping column is to reverse the absorption operation by transferring the sulfur back into the gas phase, thus regenerating the clean monoethanolamine solution. The stripping column is often packed to aid mass transfer. The sulfur as H_2S is cooled to remove water vapour. The water vapour can then be fed back into the plant to prevent the ethanolamine solution from becoming too concentrated. H_2S can then be isolated for further processing. Heat is extracted from the clean monoethanolamine emerging from the bottom of the stripping column and used to heat the H_2S-rich monoethanolamine exiting the absorption column. The clean monoethanolamine is then cooled further by heat exchange with air or water and fed back into the top of the absorber, thus completing the unit operation.[2] This alkanolamine process is also suitable for the absorption of CO_2 from feedstocks.

Absorption in Ammonia Solution

Alkanolamine processes are only really suitable for the purification of gas feedstocks that contain H_2S and CO_2 as the only impurities. They cannot be used for the purification of coal gas, for example, which contains COS, CS_2, HCN, pyridine bases, thiophene, mercaptans, ammonia and traces of nitric oxide in addition to CO_2 and H_2S impurities, since the alkanolamine will either react with these impurities or form non-recoverable residues.[2] H_2S can be removed from coal gas feedstocks by absorption in aqueous ammonia at room temperature using an absorption column similar to that described above for alkanolamine absorption. The reactions involved are:

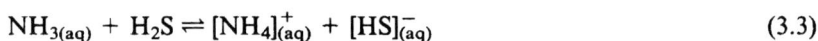

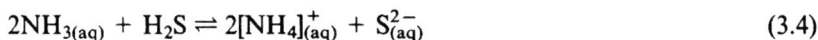

$$NH_{3(aq)} + H_2S \rightleftharpoons [NH_4]^+_{(aq)} + [HS]^-_{(aq)} \qquad (3.3)$$

$$2NH_{3(aq)} + H_2S \rightleftharpoons 2[NH_4]^+_{(aq)} + S^{2-}_{(aq)} \qquad (3.4)$$

H_2S was generally found to be present as $[HS]^-$ ions in solution. The sulfide concentration (S^{2-}) was negligible except at very high pH (> 12).[2]

Absorption in Alkaline Salt Solutions

Alkaline salt solutions formed from sodium or potassium and a weak acid anion such as carbonate or phosphate have been used in regenerative H_2S removal. They can be used to absorb H_2S, CO_2 and other acid gases. The weak acid acts as a buffer, preventing the pH from changing too rapidly on absorption of the gases.

The Seaboard process was introduced in 1920 by the Koppers Company and was the first regenerative liquid process for H_2S removal which was developed for large scale industrial use.[2] In this process, H_2S is absorbed at room temperature into a solution of dilute sodium carbonate in an absorber column similar to the one described for monoethanolamine. The gas stream containing the H_2S flows countercurrently to the sodium carbonate solution in the absorber column. The reaction can be represented as:

$$Na_2CO_3 + H_2S \rightleftharpoons NaHCO_3 + NaHS \tag{3.5}$$

The H_2S is then regenerated in a separate column using a countercurrent flow of low pressure air. The regenerated H_2S is oxidised by combustion to SO_2 and vented into the atmosphere. 85–95% of the H_2S can be removed from the gas stream by this process. It has the advantage that it is simple and economical to operate, but the disposal of the extracted sulfide as sulfur dioxide is environmentally damaging. The oxidation step is also accompanied by further oxidation of some of the H_2S to thiosulfate which is retained in the stripping column and hence contaminates recycled sodium carbonate, lowering its H_2S absorption capacity. Any HCN which may be present in the feedstock, especially if it is coal gas, is also absorbed by the Na_2CO_3 solution and is oxidised in air in the stripping column to NaSCN by reacting with NaHS, *i.e.*:

$$2NaHS + 2HCN + O_2 \rightarrow 2NaSCN + 2H_2O \tag{3.6}$$

Koppers Company later developed a vacuum distillation process which allowed the H_2S to be recovered rather than combusted by operating at low pressure and using steam as the stripping vapour.[2]

A number of processes have also been developed for absorption at elevated temperature. One of these is the Benfield process.[2] This process uses hot potassium carbonate and can remove 90% of the H_2S from a feedstock and, to a lesser degree, carbonyl sulfide, carbon disulfide, and mercaptans. It can also be used to remove CO_2, SO_2 and HCN. The Benfield process is still used today, and there are several hundred plants located throughout the world. The reactions involved are:

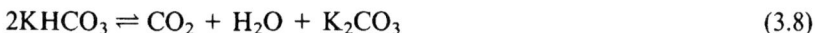

$$K_2CO_3 + H_2S \rightleftharpoons KHCO_3 + KHS \tag{3.7}$$

$$2KHCO_3 \rightleftharpoons CO_2 + H_2O + K_2CO_3 \tag{3.8}$$

It is essential to have some CO_2 in the gas stream to prevent the KHS from being converted to K_2S which is non-regenerable, *i.e.*:

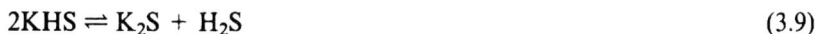

$$2KHS \rightleftharpoons K_2S + H_2S \tag{3.9}$$

Carbonyl sulfide present in the feedstock is readily hydrolysed to H_2S and CO_2:

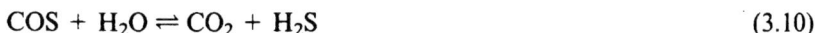

$$COS + H_2O \rightleftharpoons CO_2 + H_2S \tag{3.10}$$

Carbon disulfide is first hydrolysed to carbonyl sulfide:

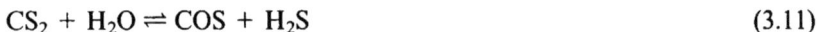

$$CS_2 + H_2O \rightleftharpoons COS + H_2S \qquad\qquad (3.11)$$

and then further hydrolysed as in equation 3.10. The two stage hydrolysis makes carbon disulfide removal considerably less efficient than that of carbonyl sulfide.[2] The Benfield process has been modified further in recent years and it has been reported that it can reduce H_2S levels in the exit gas to < 1 ppm. This is accomplished by using two hot K_2CO_3 solutions of different concentrations and carrying out the purification step in two stages.

3   Removal of H_2S by Oxidation

Claus Process

The Claus process is used to recover sulfur from feedstocks containing high concentrations (*ca.* 1–4 wt%) of H_2S. In this reaction, some of the H_2S is oxidised in air to SO_2 at *ca.* 550 °C.

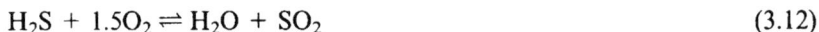

$$H_2S + 1.5O_2 \rightleftharpoons H_2O + SO_2 \qquad\qquad (3.12)$$

The remaining H_2S then reacts with the SO_2 at *ca.* 400 °C over an alumina catalyst forming sulfur.

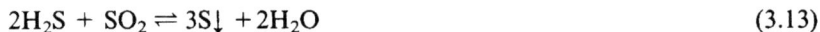

$$2H_2S + SO_2 \rightleftharpoons 3S\!\downarrow + 2H_2O \qquad\qquad (3.13)$$

This method cannot be used to reduce the sulfur content of gas feedstocks to ppm levels as equation 3.13 is equilibrium limited.[2] Further complications arise from the reaction of H_2S with CO_2 present in the feedstock which form COS and CS_2 in the high temperature oxidation stage of the process. The alumina catalyst slowly becomes deactivated by deposition of carbonaceous material from hydrocarbons present in the feedstock and sulfate formed by overoxidation of H_2S. The catalyst can be regenerated by oxidising these carbonaceous deposits.[2]

Superclaus Process

A recent extension of the Claus process is the development of the Superclaus process.[4] The bulk of the H_2S is removed in a conventional Claus process, then the remainder is removed by selective oxidation in oxygen to sulfur on a catalyst comprised of a mixture of iron and chromium oxides supported on α-Al_2O_3. The alumina support has a low surface area and wide pores. The catalyst prevents overoxidation of the sulfur to SO_2, it is insensitive to high water vapour concentration and excess O_2, and it is highly selective for H_2S oxidation (other compounds in the process gas such as COS, CS_2, H_2 and CO are not oxidised). In this manner, this process can be used to increase sulfur recovery to $> 99\%$.

Iron Oxide Process

This process was introduced in England in the 19th century, and is used in coal gas purification.[2,5,6] H_2S is allowed to react with hydrated iron(III) oxide at room temperature, to form hydrated iron(III) sulfide. The absorbent is exposed to air whereupon the sulfide is oxidised to elemental sulfur and ferric oxide is reformed. The regenerated oxide can then be reused to react with more H_2S. After several sulfiding/oxidation cycles the iron oxide becomes deactivated due to the surface becoming covered with elemental sulfur and the material must be discarded. The deposited sulfur can either be burned to SO_2 and used for sulfuric acid manufacture, or recovered by solvent extraction and recrystallisation. The reaction takes place at *ca.* 55 °C under alkaline conditions since the hydrated ferric sulfide loses its water of crystallisation and is converted to FeS_2 and Fe_8S_9 in acid or neutral conditions. FeS_2 and Fe_8S_9 are converted to iron(II) sulfate and polysulfides so that the hydrated iron(III) oxide cannot then be regenerated.

The hydrated iron(III) oxide is either used unsupported, supported on wood shavings, or as natural bog ore which is comprised of hydrated iron oxide combined with peat. Sodium hydroxide is added to maintain the alkalinity.

An extension of this process is the use of iron oxide suspensions in aqueous Na_2CO_3 in which the H_2S reacts with Na_2CO_3 at room temperature.[2] The sodium hydrogensulfide and sodium hydrogencarbonate formed in this process react with hydrated iron(III) oxide to form the sulfide.

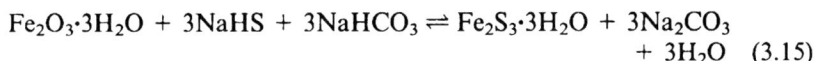

$$H_2S + Na_2CO_3 \rightleftharpoons NaHS + NaHCO_3 \qquad (3.14)$$

$$Fe_2O_3 \cdot 3H_2O + 3NaHS + 3NaHCO_3 \rightleftharpoons Fe_2S_3 \cdot 3H_2O + 3Na_2CO_3 + 3H_2O \qquad (3.15)$$

The hydrated iron oxide can be regenerated by oxidation.

$$2Fe_2S_3 \cdot 3H_2O + 3O_2 \rightleftharpoons 2Fe_2O_3 \cdot 3H_2O + 6S \qquad (3.16)$$

Most of the sulfide is precipitated out as elemental sulfur, but some is further oxidised to thiosulfate.

Activated Carbon Process

This process uses activated carbon as a catalyst to promote the oxidation of H_2S to sulfur at ambient temperatures.[2] The sulfur can then be extracted using ammonium sulfide and the carbon reused. It has a limited lifetime, however, due to the deposition of tar and polymeric materials on the surfaces of the carbon particles.

4 Removal of H₂S and Other Sulfur Compounds Using Solid Sorbents

High Temperature Sorbents

Most of the work carried out to date on the development of absorbents for H_2S removal has been on materials that are used at temperatures in excess of 300 °C. Westmoreland and Harrison investigated the thermodynamic feasibility of using various metal oxides for the desulfurisation of gas feedstocks and identified the oxides of Fe, Mo, Zn, Mn, Ca, Ba, Sr, Cu, W, Co and V as being suitable for desulfurisation when used at temperatures greater than 300 °C.[7] Westmoreland *et al.* also studied the initial rates of reaction between H_2S and MnO, CaO, ZnO and V_2O_3 over the temperature range 300–800 °C using a thermobalance reactor.[8] The hydrogen sulfide concentration in the feed was varied from 1.9 to 7.0% by volume. All the reactions were first order with respect to H_2S and obeyed the Arrhenius equation. Further studies were then carried out with single pellets of ZnO and 1–4 mol% H_2S in which sulfidation was allowed to proceed to completion.[9] Under these conditions mass transfer and diffusion effects would be observed in addition to adsorption and absorption, whereas mass transfer effects were unimportant in measurements of the initial rates carried out previously. The reaction was rapid and went to completion over the temperature range 600–700 °C.

High temperature H_2S absorption studies have also been carried out using zinc titanate.[10] The rate of formation of sulfide and saturation coverage of sulfide on ZnO and zinc titanate were similar at 720 °C, but the sulfide coverage on TiO_2 was much lower. Jalan's group studied the initial rate of sulfidation of ZnO at 650 °C and found that the uptake was low.[11,12] He attributed this to pore blocking and sintering of the ZnO at this reaction temperature. The ZnO could be regenerated using a mixture of steam and air at 650 °C but the pore structure was damaged by this treatment. Yumuru and Furimsky looked at the reaction of H_2S with iron(III) oxide, calcium oxide and zinc oxide over the temperature range 600–800 °C.[13] A feed gas of 10% H_2S in nitrogen was used, and H_2S, H_2 and SO_2 concentrations exiting from the sorbent bed were determined with respect to time. The sulfur uptake/gram of sorbent at 600 and 700 °C followed the sequence Fe_2O_3 > CaO > ZnO, whereas the sulfur uptake per unit surface area followed the order ZnO > Fe_2O_3 > CaO. Each oxide interacted with H_2S differently. H_2S decomposed during adsorption as well as reacting to form the sulfide in the case of CaO, indicating that the decomposition of H_2S may be catalysed by CaO. The predominant reaction of H_2S with ZnO was the formation of ZnS, but some decomposition was observed as the reaction proceeded, indicating that in this case ZnS may catalyse the decomposition of H_2S. Sulfur dioxide was detected in the exit gas in the early stages of the reaction of Fe_2O_3 with H_2S. A redox reaction route was proposed in which the iron(III) oxide was converted to iron(II) oxide.

$$H_2S \rightarrow H_2 + \tfrac{1}{2}S_2 \tag{3.17}$$

$$2Fe_2O_3 + H_2S \rightarrow 4FeO + H_2 + SO_2 \tag{3.18}$$

$$2Fe_2O_3 + \tfrac{1}{2}S_2 \rightarrow 4FeO + SO_2 \tag{3.19}$$

$$Fe_2O_3 + H_2 \rightarrow 2FeO + H_2O \tag{3.20}$$

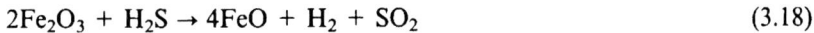

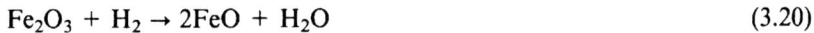

The removal of H_2S by decomposition has also been described using bimetallic supported ruthenium catalysts.[14] The catalysed decomposition was carried out in a fluidised bed reactor over the temperature range 610–730 °C and at a pressure of 160 kPa. The best catalyst was comprised of 0.5% MoS_2/0.5% RuS_2/γ-Al_2O_3. This was a very clean reaction, as the sulfur was recovered for sale, and the hydrogen could be recycled for use in hydrodesulfurisation or other hydrotreating processes.

Low Temperature Sorbents

Comparatively little work has been carried out on sorbents for low temperature desulfurisation. The main advantage of removing H_2S at low temperatures is that processing costs are substantially reduced. The use of hydrated iron oxides for H_2S removal at ambient temperatures has already been discussed (see iron oxide process). These absorbents are extremely efficient, but iron sulfide is pyrophoric in air and has therefore recently been replaced by zinc oxides in Europe.[15,16] ZnO absorbs H_2S stoichiometrically at 350 °C, but the rate of reaction decreases rapidly as the temperature is lowered. The rate of reaction is strongly dependent on the morphology as well as the metal oxide used, and this has been demonstrated by the recent development of a high surface area ZnO with an improved H_2S uptake.[17]

Stirling et al.[18] investigated the initial rate of reaction of H_2S with ZnO and ZnO doped with ca. 5 wt% CuO, Fe_2O_3 or Co_3O_4. The doped materials were prepared by one of two routes: (i) impregnation of ZnO with copper(II), iron(III) or cobalt(II) nitrate salts; (ii) precipitation of the zinc oxide or doped zinc oxides from their nitrate salts with base. These precursors to the oxides were then calcined at 350 °C for 16 hours to form the oxides. The extent of reaction (calculated from the ratio of the number of moles of H_2S absorbed to the number of moles of H_2S required for complete conversion of the oxide to sulfide) was higher for all the absorbents prepared by a precipitation route and this could at least partly be attributed to the fact that higher surface areas were obtained for the materials prepared by the precipitation rather than the impregnation route. The best absorbent was the 5 wt% CuO/ZnO prepared by the precipitation route. The H_2S absorption was found to be proportional to surface area, and it was concluded that the main role of the dopant was to increase the surface area of the zinc oxide.

Davidson, Lawrie and Sohail studied the rates of reaction of H_2S in nitrogen over the concentration range 0.05–0.8% H_2S in N_2 with ZnO and a series of doped zinc oxides at temperatures of 0–45 °C.[19] The initial rates varied considerably, and this could be attributed to differences in the morphology of the ZnO particles induced by doping. They identified three stages for the reaction of H_2S with the ZnO and doped zinc oxides. In the first stage, the rate of reaction was fast. The H_2S diffused to the ZnO surface and reacted to form a surface sulfide and water. A second fast reaction stage then occurred in which the rate of reaction was controlled by the diffusion of water out of the pellet. During this process, water was adsorbed and desorbed continuously within the body of the solid. At high conversions, a pseudo steady state conversion of the oxide was established. This proceeded at a slower rate and was thought to involve the diffusion of Zn^{2+} to the surface to react with the sulfide and H^+ from the adsorbed H_2S to hydroxylated zinc oxide forming water.[20] The hydroxylated surface was formed by a hydration step preceding H_2S adsorption at the pseudo steady state conversion. It was thought that this cation diffusion mechanism was more likely to occur than the transport of bulk ZnO to the reaction interface which would be difficult at the low temperatures at which this reaction was studied. The reaction was strongly promoted by water and independent of zinc oxide conversion and H_2S partial pressure at pseudo steady state conversion.

A series of mixed Co/Zn oxides have been prepared by a precipitation route and used to measure the initial rates of absorption of H_2S from a 2% H_2S in nitrogen feedstock at 30 °C.[21,22] The H_2S absorption capacity was found to increase with increase in cobalt concentration, with the reaction being virtually stoichiometric when pure Co_3O_4 was used. Cognion[23] used a 5% CuO/ZnO absorbent over the temperature range 200–370 °C for the desulfurisation of a feedstock containing H_2S and some organosulfides. The copper was found to absorb H_2S and, in the presence of hydrogen, it hydrogenated some sulfur compounds. The copper also acted as a promoter for ZnO, enabling it to remove less active organosulfides such as CS_2 by exchange reactions, *i.e.*:

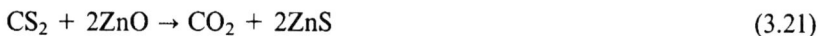

$$CS_2 + 2ZnO \rightarrow CO_2 + 2ZnS \tag{3.21}$$

Holliman *et al.* synthesised feroxyhyte (δ'-FeOOH) and used it to measure the initial rate of H_2S uptake from 2% H_2S/N_2.[24] 80% of the feroxyhyte was converted to iron(II) sulfide. On initially exposing the sample to air at room temperature, an amorphous hydrated iron oxide species formed and the sulfide was oxidised to sulfur. After several days, the amorphous oxide recrystallised as goethite (α-FeOOH).

A method has been developed for the simultaneous removal of COS and H_2S from coke oven gas.[25] α-FeOOH is partially dehydrated at *ca.* 200 °C and then used for desulfurisation at 40–60 °C in a stream of 50–100 ppm of H_2S and/or 50–100 ppm of COS. The COS is hydrolysed to H_2S and this is stabilised as iron sulfide on the surface of the iron oxide:

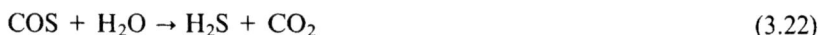

$$COS + H_2O \rightarrow H_2S + CO_2 \tag{3.22}$$

Two active sites were identified in the partially dehydrated α-FeOOH:[25] (i) Fe_2O_3 which was active for H_2S absorption; and (ii) $Fe_2O_3 \cdot xH_2O$ which was active for COS hydrolysis and H_2S absorption. H_2S formed from the hydrolysis of COS and H_2S present in the feedstock would then react with $Fe_2O_3 \cdot xH_2O$:

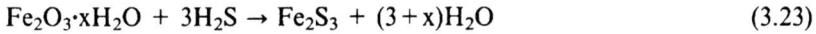

$$Fe_2O_3 \cdot xH_2O + 3H_2S \rightarrow Fe_2S_3 + (3+x)H_2O \qquad (3.23)$$

H_2S would also react at the Fe_2O_3 site:

$$Fe_2O_3 + 3H_2S \rightarrow 2FeS + 3H_2O + S \qquad (3.24)$$

The main reaction occurring at this site would however be:

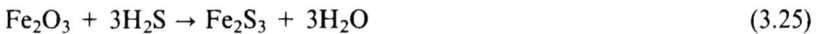

$$Fe_2O_3 + 3H_2S \rightarrow Fe_2S_3 + 3H_2O \qquad (3.25)$$

A different approach to low temperature H_2S removal is to use an adsorbent to 'concentrate' the H_2S and then to desorb and combust it.[26] A zeolite molecular sieve 13X was found to be effective in removing H_2S from a feedstream of 10 ppm H_2S in helium at 25 °C. Activated carbon was found to be a better adsorbent if the feedstream also contained 0.9% water. The adsorbed sulfide was then oxidised to sulfur in the presence of oxygen. Low temperature (25 °C) oxidation over the activated carbon can be carried out in the presence of oxygen provided that only small amounts of H_2S are involved and that the feedstream contains some moisture. The reaction was thought to take place in the water that had condensed in the pores. It was proposed that the H_2S, present as HS^- in solution in the pores, would be oxidised to sulfur using dissociatively adsorbed oxygen on the carbon. In the absence of moisture and/or at high concentrations of H_2S, the sulfide can be oxidised over the temperature range 150–250 °C to SO_2 and SO_3 using a monolith supported platinum or palladium catalyst. These high temperature oxidation experiments were carried out using a feed gas of 1.8% methane, 21% oxygen, 20 ppm H_2S and the balance as nitrogen in order to simulate emissions from sewerage gases which contain methane. The methane would not be oxidised at these temperatures, but it is important to study the interaction of H_2S with sorbents and catalysts in the presence of all the gases likely to be present in a feedstock in order to obtain a realistic assessment of sorption capacity/reactivity in industrial conditions. With the continuing worldwide interest in emission control, it is likely that this will be a fruitful area of research for many years to come.

5 Conclusions

Although much progress has been made in the development of sorbents that can remove H_2S at low temperatures (and indeed attention is now also focusing on the interactions of gases in these materials), the mechanisms for the low temperature diffusion of ions through the solids are still poorly understood and differ for many of the sorbents studied. Generally, sorption is accompanied by reaction with the conversion, for example, of a metal oxide to a metal sulfide and water.

In broad terms, the interaction of adsorbed H_2S with the solid can proceed by either a pore or a lattice diffusion mechanism. H_2S is dissociatively adsorbed on the surface of the solid, forming H^+ and HS^-. If pore diffusion predominates, the sorption capacity will be limited by the rate of diffusion of HS^- to fresh surface (oxide) sites in the pores. Pore size and pore accessibility as the reaction proceeds will therefore be important. Pore diffusion is likely to be rate limiting if the sorbent particles are large (*ca.* 2.5–3.5 mm) since it will take longer for the H_2S to diffuse into the centre of a particle.[12] Pore diffusion has been found to be rate limiting in high temperature studies of the reaction of H_2S with zinc oxide.[9]

In lattice diffusion, HS^- will react with oxide at the surface to form a surface sulfide and water. The reaction can then be controlled by one or more diffusion routes. The oxide from the bulk of the lattice can replenish the depleted surface oxide and water can diffuse to the surface, and/or HS^- can diffuse into the oxide lattice. The reaction will cease either when all the oxide has reacted or when an inert sulfide shell forms which limits the diffusion process. In reality, bulk diffusion of anions is unlikely at the low temperatures used in these studies and surface reconstruction may occur on reaction. Thus, measurements of the reaction of H_2S (0.05–0.8% in N_2) with ZnO over the temperature range 0–45 °C taken when the reaction was operating at pseudo steady state conversion, indicated that a water promoted cation diffusion mechanism operated at these low temperatures[19] (see previous discussion of low temperature sorbents). The occurrence of reconstruction on reaction is illustrated by the studies of the initial rates of reaction of H_2S with the mixed Co_3O_4/ZnO materials discussed there. Reconstruction and oxide segregation occurred in these mixed oxide materials when they were used as sorbents for H_2S.[22] The spinel $ZnCo_2O_4$, which was present at the surface of these mixed oxides, oxidised the H_2S to sulfur. Both Co(II) and Co(III) were detected at the surface by XPS after sulfidation whereas only Co(III) was present initially (*i.e.* in the surface spinel $ZnCo_2O_4$). It was proposed that the oxides segregated in the presence of H_2S and were then sulfided. The following equations are consistent with the observed XPS data:

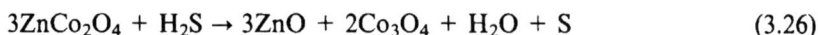

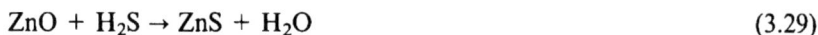

$$3ZnCo_2O_4 + H_2S \rightarrow 3ZnO + 2Co_3O_4 + H_2O + S \qquad (3.26)$$

$$Co_3O_4 + H_2S \rightarrow 3CoO + H_2O + S \qquad (3.27)$$

$$CoO + H_2S \rightarrow CoS + H_2O \qquad (3.28)$$

$$ZnO + H_2S \rightarrow ZnS + H_2O \qquad (3.29)$$

Interestingly, analysis of the sulfided samples by transmission electron microscopy revealed the presence of microcrystalline membraneous sheets which linked the particles together. Micrographs of the oxides before and after sulfiding are shown in Figure 5.14 (see Chapter 5). The sheets contained cobalt, zinc and sulfur and analysis by electron energy loss spectroscopy

(EELS) showed that they were zinc rich. The surface segregation and formation of the membraneous sheets was attributed to surface reconstruction occurring on sulfiding the mixed oxides.

Several theoretical models have been proposed for the reaction of solids with gases. The two most common models are the shrinking core and the grain model.[27,28] The shrinking core model assumes that the H_2S diffuses through a gas film which surrounds a spherical particle of sorbent to the oxide surface where it reacts. The oxide 'core' then shrinks as the reaction proceeds. The H_2S must continually diffuse through the sulfided sorbent to unreacted oxide. The rate-controlling step could be the diffusion of H_2S through the gas film, the diffusion of H_2S through the sulfided sorbent or the reaction of H_2S with the oxide at the sulfide/oxide interface, but it is generally found that the gas film does not offer much resistance to mass transfer and so one of the latter two steps are rate limiting.[28] The grain model more effectively represents the sulfidation reaction and has been used to model the kinetics of the reaction of H_2S with ZnO[9] and zinc ferrites[28] at high temperatures. In the grain model, it has been assumed that each sorbent particle is spherical and that it contains cylindrical pores. The diffusion of the H_2S through the pores of the particle was assumed to be much greater than that through the sulfide film which formed at the reaction interface. The reaction takes place at the sulfide/oxide interface and is assumed to be first order. The model allows the surface area, porosity and diffusion coefficients to be varied during the sulfidation process so that it is in accord with changes in the material as the reaction proceeds. The grain model correctly predicted the observed changes in concentration of H_2S with respect to time for the reaction of H_2S with zinc ferrites at 540 °C[28] and for the reaction of H_2S with ZnO over the temperature range 600–700 °C.[9] Sorption kinetics are discussed in more detail in Chapter 7.

6   References

1 C.N. Satterfield, 'Heterogeneous Catalysis in Industrial Practice', 2nd edn, McGraw Hill, New York, 1991.
2 A. Kohl and F. Riesenfeld, 'Gas Purification', 4th edn, Gulf Publishing Company, Houston, 1985, and references therein.
3 J.R. Fair, D.E. Steinmeyer, W.R. Penney and B.B. Crocker, in 'Perry's Chemical Engineers' Handbook', ed. R.H. Perry, D.W. Green and J.O. Maloney, 7th edn, McGraw Hill, USA, 1997, Section 14.
4 V. Nisselrooy and J.A. Lagas, *Catal. Today*, 1993, **16**(2), 263.
5 J.E. Garside and R.F. Phillips, in 'A Textbook of Pure and Applied Chemistry', ed. S.C. Laws, Pitman and Sons, London, 1962, p. 590.
6 G.U. Hopton and R.H. Griffith, *Gas J.*, 1946, **247**, 4311.
7 P.R. Westmoreland and D.P. Harrison, *Environ. Sci. Technol.*, 1976, **10**(7), 659.
8 P.R. Westmoreland, J.B. Gibson and D.P. Harrison, *Environ. Sci. Technol.*, 1977, **11**(5), 488.
9 J.B. Gibson and D.P. Harrison, *Ind. Eng. Chem., Process Dev.*, 1980, **19**(2), 231.
10 R.V. Siriwardani and J.A. Poston, *Appl. Surf. Sci.*, 1990, **45**, 131.
11 V. Jalan, *Int. Gas Res. Conf.*, 1981, 291.
12 C. Lawrie, 'A Study of the Reaction Between H_2S and ZnO', Ph.D. Thesis, University of Edinburgh, 1990.

13 M. Yumura and E. Furimsky, *Ind. Eng. Chem., Process Dev.*, 1985, **24**, 1165.
14 D. Cao and A.A. Adesina, *Catal. Today*, 1999, **49**, 23.
15 K. Eddington and P. Carnell, *Oil Gas J.*, 1991, **89**, 69.
16 G.W. Spicer and C. Woodward, *Oil Gas J.*, 1991, **89**, 76.
17 P.J.H. Carnell and P.E. Starkey, *Chem. Eng.*, 1984, **408**, 30.
18 T. Baird, P.J. Denny, R.W. Hoyle, F. McMonagle, D. Stirling and J. Tweedy, *J. Chem. Soc., Faraday Trans.*, 1992, **88**(22), 3375.
19 J.M. Davidson, C.H. Lawrie and K. Sohail, *Ind. Eng. Chem. Res.*, 1995, **34**(9), 2981.
20 J.M. Davidson and K. Sohail, *Ind. Eng. Chem. Res.*, 1995, **34**(11), 3675.
21 T. Baird, K.C. Campbell, P.J. Holliman, R.W. Hoyle, D. Stirling, B.P. Williams and M. Morris, *J. Mater. Chem.*, 1997, **7**(2), 319.
22 T. Baird, K.C. Campbell, P.J. Holliman, R.W. Hoyle, M. Huxam, D. Stirling, B.P. Williams and M. Morris, *J. Mater. Chem.*, 1999, **9**(2), 599.
23 J.M. Cognion, *Chim. Ind. Gen. Chim.*, 1972, **105**, 757.
24 T. Baird, K.C. Campbell, P.J. Holliman, R.W. Hoyle, D. Stirling and B.P. Williams, *J. Chem. Soc., Faraday Trans.*, 1996, **92**(3), 445.
25 K. Miura, K. Mae, T. Inoue, T. Yoshimi, H. Nakagawa and K. Hashimoto, *Ind. Eng. Chem. Res.*, 1992, **31**, 415.
26 V. Meeyoo, J.H. Lee, D.L. Trimm and N.W. Cant, *Catal. Today*, 1998, **44**(1–4), 67.
27 O. Levenspiel, 'Chemical Reaction Engineering', 3rd edn, John Wiley and Sons, New York, 1999, p. 568.
28 M. Pineda, J.M. Palacios, E. Garcia, C. Cilleruelo and J.V. Ibarra, *Fuel*, 1997, **76**(7), 567.

CHAPTER 4

Clean-up of Sulfur Dioxide

1 Introduction

Despite substantial clean up of feedstocks before processing, stringent legislation to control SO_2 emissions and hence limit acid rain make SO_2 removal prior to discharge of emissions into the atmosphere essential. Flue gases from the combustion of coal generally contain $< 0.5\%$ SO_2. The largest source of SO_2, however, is from stack gases from smelters handling sulfur ores which may have up to 8% SO_2 in them.[1] Generally, some SO_3 is found in addition to the SO_2 in the combustion gases. SO_2 is more stable in excess air provided the temperature is kept above *ca.* 730 °C, but SO_3 can be produced catalytically by traces of vanadium which is frequently found in oil residues[2] and iron pyrites which are present in coal. Other sources of SO_2 include emissions from car exhausts (where it is found together with nitrogen oxides), emissions from municipal refuse incineration and emissions from sulfur-containing fuel-fired boilers. A number of processes are described here for SO_2 removal from gas feedstocks. They can be categorised as absorption in liquids and sorption by solids. One of the problems is that SO_2 does not often occur in isolation, so it is necessary to design sorbents that can remove SO_2 in the presence of other gases such as nitrogen oxides. The sorbents should ideally be regenerable, as this will lower operating costs. Although a large number of processes have been proposed for removing SO_2 from gas streams, few have been used commercially. Most of the commercial processes are for flue gas desulfurisation (FGD),[3] and absorption in a lime slurry is the main process used due to the fact that lime is cheap and readily available.

2 Absorption in Liquids

Processes which involve removal of SO_2 by absorption in a liquid include absorption into a soluble alkali, lime, aqueous aluminium sulfate or sodium citrate.

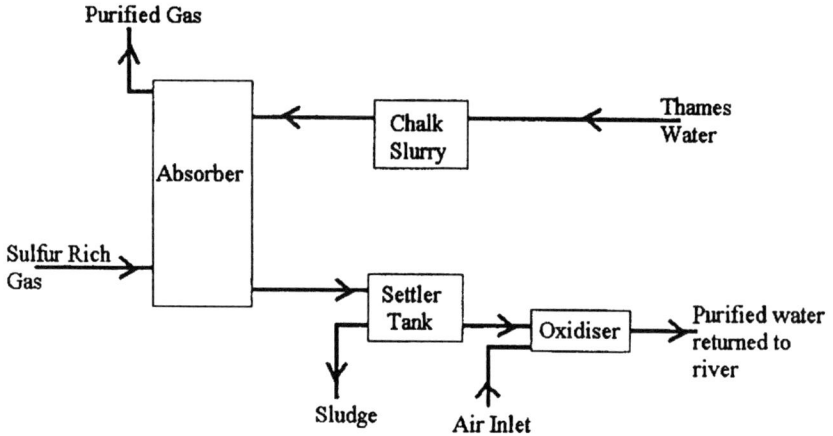

Figure 4.1 *Diagram of the Battersea process for removal of sulfur dioxide from flue gas.*
(Adapted from ref. 1, p. 310. Used by permission. Adapted from *Gas
Purification* © 1985, Gulf Publishing Company, Houston, Texas, 800-231-
6275. All rights reserved)

SO$_2$ Clean-up Using a Lime/Limestone Process

The SO$_2$ containing flue gas is passed into an aqueous slurry of lime or limestone
in this process. The SO$_2$ reacts with the lime/limestone to form calcium sulfite
and calcium sulfate which can be collected for disposal. The purified gas is then
discharged to the atmosphere. The process was originally formulated by
Eschellman, and the earliest commercial application was at Battersea Power
Station in London in 1931. A schematic diagram of the process is shown in
Figure 4.1.[1]

Thames river water, which was naturally alkaline, was used and chalk was
added to the water prior to passing it into the top of an absorber column. The
sulfur-rich flue gas was passed into the bottom of the absorber at 133 °C, flowed
countercurrently to the chalk (CaCO$_3$) slurry and emerged as purified gas from
the top of the column. The river water from the bottom of the absorber was
passed into a settling tank to remove the calcium sulfite and calcium sulfate
sludge. The water was then passed into an oxidising tank. Air was bubbled into
the oxidising tank to oxidise dissolved calcium sulfite to calcium sulfate. This
reduced the amount of dissolved sulfite in the line which could be oxidised *in situ*
and become deposited on the equipment. Modern plants for SO$_2$ removal are
based on this design. In the lime/limestone process the SO$_2$ dissolves in water
and a portion of it ionises:

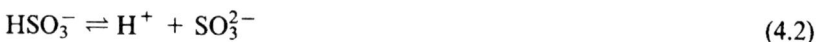

$$SO_2(aq) + H_2O \rightleftharpoons H^+ + HSO_3^- \tag{4.1}$$

$$HSO_3^- \rightleftharpoons H^+ + SO_3^{2-} \tag{4.2}$$

Lime (Ca(OH)$_2$) or limestone (CaCO$_3$) is then added to provide a source of Ca^{2+}. The lime dissolves and the limestone is dispersed in water. The limestone must be finely ground to give a reasonable rate of reaction. Ca^{2+} then reacts with SO_3^{2-}(aq) to form a precipitate of $CaSO_3 \cdot \frac{1}{2}H_2O$ and sulfate (formed from the oxidation of sulfite) to form the hydrated calcium sulfate, gypsum.

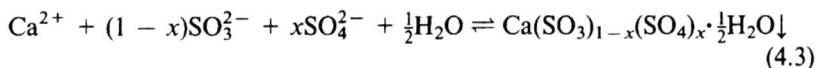

$$Ca^{2+} + (1 - x)SO_3^{2-} + xSO_4^{2-} + \tfrac{1}{2}H_2O \rightleftharpoons Ca(SO_3)_{1-x}(SO_4)_x \cdot \tfrac{1}{2}H_2O \downarrow$$

$$(4.3)$$

This process is most successful when the SO_2 concentration in the gas phase is below 3000 ppm.

Alternative Absorbents for SO_2 Removal

Other absorption processes include absorption into aqueous sodium carbonate or hydroxide to form sodium sulfite, absorption in ammonia solution to form ammonium sulfate and absorption in basic aluminium sulfate solution.[1] The basic aluminium sulfate process is known as the Dowa process, and it was developed by the Dowa Mining Company in Japan.[4] In this process, SO_2 reacts with the basic aluminium sulfate to form aluminium sulfite. This is then oxidised in air to the sulfate. Limestone is then added to reform the basic aluminium sulfate solution and remove the excess sulfate as gypsum.[1]

Absorbents for the Removal of Both SO_2 and NO_x

As already stated, SO_2 emissions are frequently accompanied by the emission of nitrogen oxides (NO_x), and ideally both gases should be removed in one step. One process that has been evaluated is the oxidation of these gases using Ce(IV) in acid.[5] The Ce(IV) flowed countercurrently to the waste gases in a column and could be regenerated at the anode of an electrochemical cell in a continuous cycle. The recovery of cerium-free nitric and sulfuric acids for further processing was also investigated.

Another absorbent based process which is being developed by Dravo Lime Company for the removal of NO_x and SO_2 from flue gas is known as the ThioNO$_x$ process.[6] The SO_2 is removed in slaked lime [Ca(OH)$_2$] to which MgO is added to increase the efficiency of SO_2 removal. The NO_x is removed by an iron(II) EDTA chelate which is added to the lime slurry. The SO_2 is oxidised to calcium sulfite and the NO binds to the iron-EDTA chelate. Pilot plant trials showed that up to 60% of the NO_x and greater than 99.5% of the SO_2 could be removed from the flue gas. Bench scale studies showed that the iron(II) EDTA solution could be regenerated electrochemically.

3 Sorption in Solids

The use of solid sorbents for the removal of SO_2 from feedstocks such as flue gases offers many processing advantages. The plant will be less complex and thus cheaper to construct, requiring less maintenance and simplifying operational procedures. Regeneration of spent sorbents will also generally be easier. Activated carbon is probably one of the most widely used sorbents for SO_2 recovery.[1,7] A copper oxide based process has also been developed by Shell.[8] As in the case of the absorbents, attention is now focusing on the development of sorbents for removal of both SO_2 and NO_x and analysis of the interaction of these gases with the sorbent has provided an insight into how the sorbents function. As in the case of sorbents for H_2S removal, however, further fundamental studies are needed to investigate the interactions of SO_2, NO_x and other emissions such as CO_2 and water in order to design sorbents that are effective for a wide range of applications. A representative sample of some of the sorbents that are either being developed or are currently in use is described here.

Activated Carbon Process

Activated carbon has been used commercially for SO_2 recovery.[1,7] It is an example of a process in which SO_2 can also be removed from gas effluent by adsorption onto a solid without reaction. This has the advantage that regeneration of the sorbent should be much easier. The carbon catalyses the oxidation of adsorbed SO_2 in excess oxygen at low temperature (*ca.* 110–180 °C). Water is required for the reaction to proceed at a reasonable rate.

$$SO_2 + \tfrac{1}{2}O_2 + H_2O \xrightarrow{\text{activated carbon}} H_2SO_4 \qquad (4.4)$$

The adsorbent can be regenerated in one of two ways: (i) washing with water to remove the sulfate as dilute sulfuric acid; (ii) heating the adsorbent to *ca.* 420 °C to reduce the sulfate to SO_2. In (ii) the carbon acts as the reducing agent, *i.e.*:

$$2H_2SO_4 + C \rightarrow 2SO_2 + 2H_2O + CO_2 \qquad (4.5)$$

An inexpensive adsorbent such as coke must be used if the adsorbent is regenerated by heating as part of the adsorbent is lost in the regeneration step.

Copper Oxide Regenerable Sorbent

CuO supported on a porous alumina has been used as a dry sorbent for SO_2 removal that can be regenerated once spent.[1] A laboratory scale process was developed by the US Bureau of Mines in *ca.* 1970.[9] The SO_2 reacts with the CuO at *ca.* 400 – 450 °C in air to form copper sulfate.

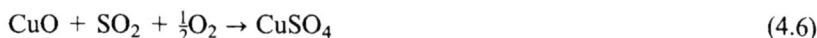

$$CuO + SO_2 + \tfrac{1}{2}O_2 \rightarrow CuSO_4 \qquad (4.6)$$

Regeneration can be carried out by treatment in reducing gases. Methane was found to be preferable to hydrogen as a reduction gas since some over reduction of the sulfate to sulfide occurred in hydrogen at low temperatures.

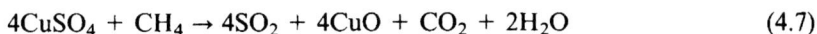

$$4CuSO_4 + CH_4 \rightarrow 4SO_2 + 4CuO + CO_2 + 2H_2O \qquad (4.7)$$

The major problem with operating this system commercially is that the rate of uptake of SO_2 on CuO is slow so a large reactor would be required on scale up. Regeneration would also be expensive.

A copper oxide based process has also been developed by Shell.[8] This process has been developed commercially and is called the Shell Flue Gas Desulfurisation Process (SFGD). It has the advantage that the absorption and regeneration steps are carried out in the same vessel at *ca.* 400 °C.

Alkali Salt Promoted $CuO/\gamma Al_2O_3$

The sorption capacity for SO_2 has been determined for 3×3 mm alumina pellets impregnated with $Cu(NO_3)_2 \cdot 3H_2O$ with and without an alkali salt to give 8 wt% CuO or 8 wt% CuO and 5 wt% LiCl, NaCl, KCl, LiBr, LiF or NaF after calcination at 600 °C in air.[10] The sorption capacities of the materials were determined at 500 °C using a feed of 1.5 vol% SO_2 in air at a total flow rate of 0.9 l/min. The alkali salt promoters both lowered the temperature at which bulk sulfation of the sorbents occurred and increased the SO_2 sorption capacity of the CuO/Al_2O_3 sorbent. The best promoter was LiCl which increased the sorption capacity of CuO/Al_2O_3 threefold after treatment with SO_2 for 150 min at 500 °C and bulk sulfation could occur at *ca.* 420 °C compared with 500 °C in the unpromoted sorbent. This was interpreted as being due to the alkali promoted decomposition of $CuSO_4$ formed by the the sulfation of the CuO in air. The decomposition reaction results in the formation of SO_3 which can react with the Al_2O_3 to form $Al_2(SO_4)_3$ at the reaction temperature.

$$CuSO_4 \rightleftharpoons CuO + SO_3 \qquad (4.8)$$

$$Al_2O_3 + 3SO_3 \rightarrow Al_2(SO_4)_3 \qquad (4.9)$$

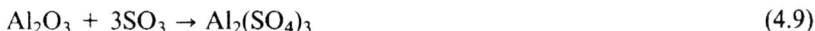

The CuO is then available for further reaction with SO_2. After sulfation, the sorbent could be regenerated by reduction in 5% H_2 in N_2 at 500 °C. The LiCl promoted material retained its enhanced sorption capacity compared with the unpromoted supported CuO for up to 30 sulfation and regeneration cycles.[10]

Removal of NO_x and SO_2

NO_x and SO_2 can be removed using a Co/Mg/Al mixed oxide prepared by calcining its hydrotalcite-like precursor at 750 °C in air.[11] The mixed oxide has the molar composition $7MgO:1Al_2O_3:1CoO$. The hydrotalcite-like precursor has a layered structure similar to that of the mineral hydrotalcite. Its structure is discussed in detail in Chapter 5. The Co/Mg/Al mixed oxide was activated in H_2 at 530 °C prior to use in NO_x removal. The activated mixed oxide proved effective at removing NO_x (300–400 ppm NO in N_2) by reduction at 750 °C in the presence of propane. It was proposed that the active sites for NO removal were reduced cobalt species and their formation was favoured at high temperatures and using reducing conditions. The sorbent could be regenerated by heating in hydrogen at 530 °C for half an hour. It was necessary to add an oxidant such as cerium(IV) oxide to the Co/Mg/Al mixed oxide for the removal of SO_2 (1400 ppm) from 3% O_2 in N_2 at 750 °C. The CeO_2 was required in order to oxidise the SO_2 to SO_3. The SO_3 can then further react with the mixed oxides to form the sulfate. The sorbent could then be regenerated at 530 °C in hydrogen.[11] Although NO_x can be removed from feedstocks most effectively from a nitrogen stream, the sorbents are effective at up to 1% O_2, which should enable NO_x and SO_2 to be removed simultaneously.

A laboratory scale powder-particle fluidised bed reactor has been used to remove both SO_2 and NO_x from simulated flue gases.[12] A fluidised bed of particles can be obtained when a gas stream flows up through the bed at sufficient velocity for the individual particles to become separated and supported by the gas phase. In the powder-particle fluidised bed a combination of coarse particles 300–700 μm and fine particles *ca.* 50 μm in size are fluidised. The coarse particles are retained in the reactor whereas the fine particles are continuously fed into the bottom and removed from the top of the bed. A diagram of a typical powder-particle fluidised bed for use in the removal of SO_2 and NO_x is shown in Figure 4.2.[12]

In the laboratory reactor study, a WO_3/TiO_2 catalyst was used for the coarse particles, and either a dust sorbent from a steel plant comprised mainly of iron(II) oxide, or a copper(II) oxide sorbent, was used as the fine particles. WO_3/TiO_2 is a catalyst suitable for the removal of NO_x by reduction with NH_3 and it was found to also catalyse the oxidation of SO_2 to SO_3 in the presence of oxygen at 500–600 °C.[12] SO_2 and SO_3 was converted to sulfate on the sorbent particles. The sorption capacity for the fine particles in a feed of 500 ppm SO_2 in air and a sorbent/S ratio of three increased with increasing temperature, giving *ca.* 80% conversion at 600 °C. Greater than 90% of the NO_x could be removed from the feedstream at a NO_x concentration of 500 ppm using a NH_3/NO mole ratio of one over the temperature range 500–600 °C. Scanning electron microscopy combined with electron imaging techniques showed that the sulfate was distributed only on the outside of the CuO sorbent particles indicating that the reaction was diffusion limited, whereas the more porous steel plant dust particles were sulfated throughout. However, the CuO was more efficient at removing SO_2 from the feedstock. There would be considerable economic advantages if combined SO_2 and NO_x removal could be carried out at lower temperatures. Further studies identified a CuO on a V_2O_5/Al_2O_3 support as an active sorbent for the removal of 60–70% SO_2 from a feedstock containing 500 ppm of SO_2 over the temperature range 300–400 °C.[13]

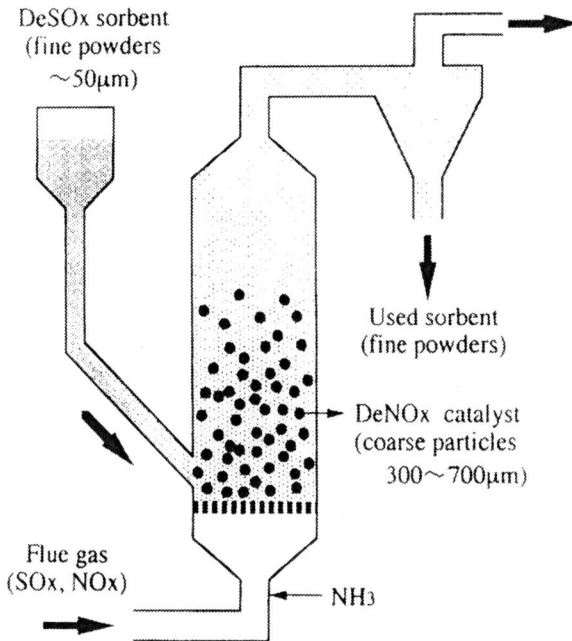

Figure 4.2 *Conceptual illustration of a powder-particle fluidised bed (simultaneous removal of SO$_2$/NO$_x$ process).*
(Reprinted from *Catal. Today*, **29**, S. Gao, N. Nakagawa, K. Kato, M. Inomata and F. Tsuchiya, p. 166. © 1996, with permission from Elsevier Science)

4 Conclusions

It has now been recognised that sulfur and nitrogen emissions cannot be considered in isolation but as contributors to the total emissions which define the global environment. Thus, for example, emissions of volatile organic compounds (VOCs) and nitrogen oxides are concentrated in urban and industrial areas and generate ozone at ground level;[3] ozone is a respiratory irritant and thus adds to the respiratory problems caused by sulfur dioxide and small soot particles originating from the VOCs (see Chapter 1). Emissions of CO$_2$ from flue gas cleaning, power plants and other sources are much greater than emissions of sulfur and nitrogen oxides, VOCs, HCFCs, *etc.* and are causing global warming. The carbon dioxide traps radiation in the earth's atmosphere causing the temperature to rise. Many of the processes used to clean up sulfur-containing emissions actually increase CO$_2$ emissions, either directly if they involve oxidation, or indirectly if energy is required for SO$_2$ extraction or sorbent regeneration.[3] Power demands are now met by oil feedstocks in many parts of the world and this limits sulfur emissions from this source as most of the sulfur is removed by hydrotreating processes. This reduction in the use of fossil fuels in power plants and also in the domestic market with the conversion of household heating to natural gas and oil-fired central heating has led to a lowering of sulfur emissions and, aided by the development of efficient absorbents and sorbents for SO_2 removal, can currently control emissions at an acceptable level. However, as oil reserves diminish we may once again return to fossil fuels as an energy source and more effective means will be needed to limit sulfur emissions. It is therefore vital that we look towards developing other energy sources with lower emissions of sulfur, nitrogen and VOC emissions, such as nuclear power and wind generators.

5   References

1 A. Kohl and F. Riesenfeld, 'Gas Purification', 4th edn, Gulf Publishing Company, Houston, 1985.

2 C.N. Satterfield, 'Heterogeneous Catalysis in Industrial Practice', 2nd edn, McGraw Hill, New York, 1991.

3 J. Ando, Chapter 26 in 'The Chemistry of the Atmosphere', ed. J.G. Calvert, Blackwell Science, Oxford, 1994, p. 363.

4 P.S. Nolan and D.O. Seaward, *Proceedings of Seminar on Flue Gas Desulfurisation*, sponsored by the Canadian Electrical Association, Ottawa, Ontario, September 1983.

5 P. Hoffmann, C. Roizard, F. Lapicque, S. Venot and A. Maire, *Process Safety and Environmental Protection*, 1997, **75**, No. B1, 43.

6 S. Tseng, M. Babu, M. Niksa and R. Coin; *Proc. 31st Intersoc. Energy Convers. Eng. Conf.*, 1996, 1956.

7 P.G. Maurin and J. Jonakin, *Chem. Eng.*, 1970, **77**, 173.

8 F.M. Doutzenberg, J.E. Naber and A.J.J. van Ginneken, 'The Shell Flue Gas Desulfurisation Process', paper presented at AIChE, 68th National Meeting, Houston, Texas, 1971.

9 D.H. McCrea, A.J. Forney and J.G. Meyers, *J. Air Pollution Control Assoc.*, 1970, 819.

10 S.M. Jeong and S.D. Kim, *Ind. Eng. Chem. Res.*, 1997, **36**, 5425.

11 A.E. Palomares, J.M. Lopez-Nieto, F.J. Lazaro, A. Lopez and A. Corma, *Appl. Catal. B: Environmental*, 1999, **20**(4), 257.

12 S. Gao, N. Nakagawa, K. Kato, M. Inomata and F. Tsuchiya, *Catal. Today*, 1996, **29**, 165.

13 S. Gao, H. Suzuki, N. Nakagawa, D. Bai and K. Kato, *Sekiyu Gakkaishi*, 1996, **39**(1), 59.

CHAPTER 5

Synthesis and Characterisation of Solid Sorbents

1 Prerequisites for Efficient Sorbents

Solid sorbents for the removal of sulfur compounds are generally single or mixed oxides synthesised either by solid state reaction or by precipitation routes. The efficiency of sorbents for removal of sulfur compounds is dependent on: (i) the surface area of the sorbent, (ii) the metal selected, (iii) the structure adopted, (iv) sorbent stoichiometry, (v) lattice and crystal defects, and (vi) the porosity of the material.

High surface areas are required for efficient sulfur uptake, and the development of a high surface area zinc oxide led to a substantial improvement in its low temperature hydrogen sulfide absorption capacity compared to that of conventional zinc oxide.[1] The surface area can be controlled by controlling the morphology and homogeneity of the precursor to the oxide absorbent. Low temperature synthesis routes are often necessary in order to obtain oxides with high surface areas or small particle sizes. However, this treatment does not necessarily give the most thermodynamically stable phase, so the materials are often amorphous, metastable, or form hydrous oxides where the water has to be removed by treatment at high temperatures.[2]

The dependence of the absorption capacity on the metal oxide and on its structure is illustrated by studies of mixed cobalt/zinc oxides.[3] The sulfur absorption capacity increased with increase in cobalt content of the mixed oxides. The increase was over and above that expected for the increased surface area at high cobalt loadings. It was probably also related to the morphology which changed from a hexagonal ZnO structure at low cobalt loadings to a cubic normal spinel structure at high cobalt loadings.

The importance of sorbent stoichiometry is illustrated by studies of the reaction of feroxyhyte (δ'-FeOOH) with H_2S.[4] Approximately 80% conversion of the feroxyhyte to iron(II) sulfide occurred at room temperature. The iron(III) in feroxyhyte was reduced to iron(II) in the form of mackinawite (FeS_{1-x}) and α-sulfur was formed.

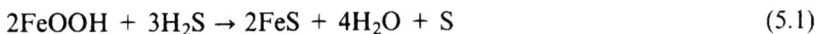

$$2FeOOH + 3H_2S \rightarrow 2FeS + 4H_2O + S \qquad (5.1)$$

A second illustration of this is the reaction of Co_3O_4 with H_2S.[3] The reaction is thought to take place in two stages: (i) reduction of Co(III) in Co_3O_4 to Co(II) oxide with the concommitant oxidation of sulfide to sulfur, and (ii) sulfiding of the cobalt (II) oxide.

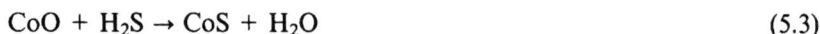

$$Co_3O_4 + H_2S \rightarrow 3CoO + H_2O + S \qquad (5.2)$$

$$CoO + H_2S \rightarrow CoS + H_2O \qquad (5.3)$$

In the presence of zinc oxide, surface reconstruction to form sheet-like structures containing zinc, cobalt and sulfur was also observed.

Gour showed that defects are beneficial in the design of efficient sorbents for sulfur removal.[5] Defects arise when atoms are displaced from the regular array in a crystal.[6,7] They can be classified as: (i) point defects which occur at single atom sites, and (ii) extended defects where the defect extends through large sections of the crystal structure. Point defects can either be intrinsic (stoichiometric) point defects in which no new ions are added to the structure, or extrinsic (non-stoichiometric) in which a foreign ion is doped into the crystal. The two main types of intrinsic point defect are the Schottky defect in which there is a pair (cation + anion) of lattice vacancies present in the crystal structure [Figure 5.1(a)] and the Frenkel point defect in which an atom has moved from a lattice position into an interstitial site [Figure 5.1(b)]. An example of an extrinsic point defect is zirconia doped with CaO. In this case some Ca^{2+} ions replace some of the Zr(IV) atoms in the crystal structure and an anion vacancy is created for each Zr(IV) replaced with Ca^{2+} to balance the charge [Figure 5.1(c)].

An example of an extended defect is an edge dislocation. Edge dislocations occur when an extra row of atoms are inserted through part of the crystal structure (Figure 5.2).

Gour studied the reaction of H_2S with ZnO. He found that the sulfur uptake could be correlated with the amount of interstitial zinc present in the sample. Interstitial zinc improved the transport of ions in the lattice.

The importance of porosity is illustrated in analysis of ZnO granules which have been used in the absorption of H_2S. Less efficient sorbents were found to have the sulfur in the outer shell of the granule or pellet, suggesting that pore diffusional resistance was rate limiting and that pore blockage may have occurred and restricted the efficiency of the usage of the pellet. The more efficient absorbents had sulfur distributed throughout the interior of the granules.[8]

2   Synthesis of Sorbents

Polycrystalline solids suitable for use as sorbents are generally synthesised by precipitation routes. This is preferable to solid state synthesis in which powdered solid starting materials are mixed and reacted directly, since high temperatures are then required for the synthesis to proceed at reasonable rates

(a)

represents an anion	represents a vacancy
represents a cation	

(b)

represents an anion	represents a vacancy
represents a cation	

(c)

represents an anion	represents a vacancy
represents a 4+ cation	
represents a 2+ cation	

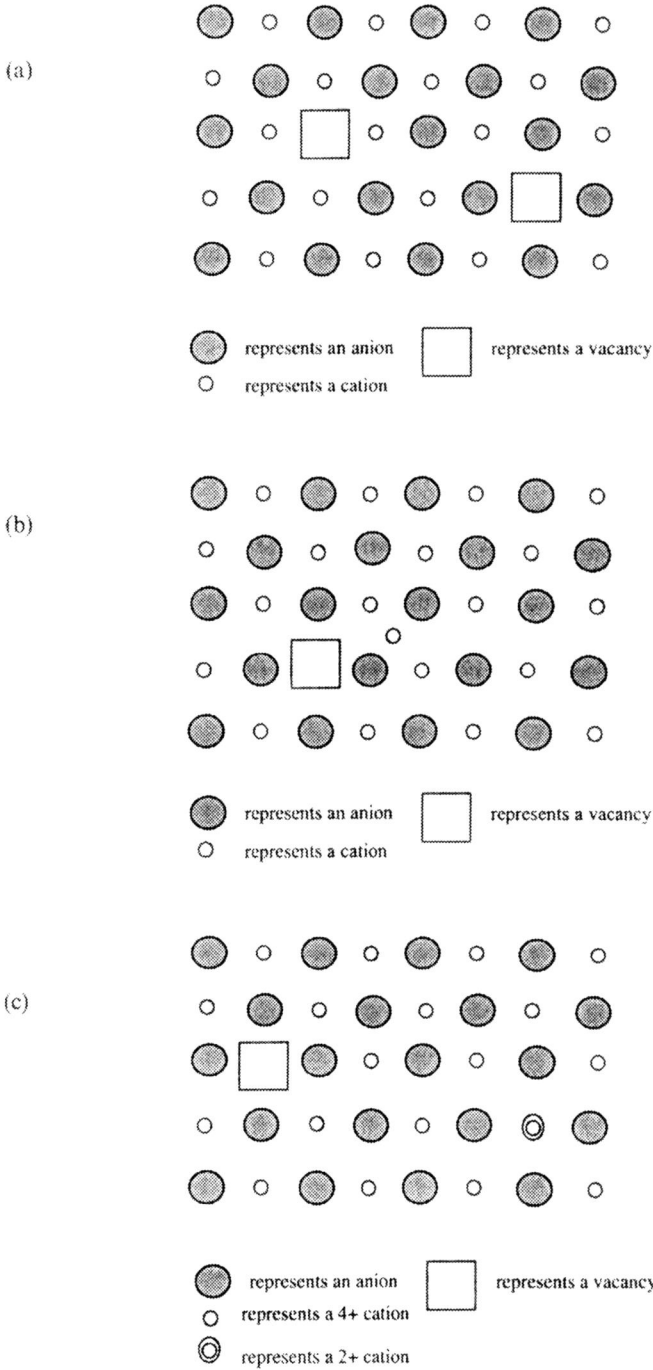

Figure 5.1 *Schematic diagram of point defects in a crystal:* (a) *Schottky defect*, (b) *Frenkel defect*, (c) *extrinsic defect*.

Figure 5.2 *Schematic diagram of an edge dislocation.*

and this results in a considerable loss in surface area. In precipitation routes, a base is added to a metal salt to precipitate the metal as a carbonate or hydroxide, for example, and this can then be decomposed to form a high surface area metal oxide. The precipitated materials are often layered structures and are often good sorbents in their own right.[9]

The precipitation process involves three stages, namely supersaturation, nucleation and crystal growth. Supersaturation occurs when a solution is in a metastable state in which it holds more dissolved solute than is required to saturate the solution. Under these conditions nucleation can occur. The solution is usually aqueous when preparing absorbents by precipitation routes, and under these conditions nucleation is always heterogeneous in that it involves deposition of solute on impurity particles such as dust present in the solvent, reagents and reaction vessel. In contrast, homogeneous nucleation occurs when impurities have been excluded and involves a spontaneous crystallisation that occurs once the build-up of supersaturation becomes high. Heterogeneous nucleation proceeds in the precipitation reaction until the degree of supersaturation reaches the metastable limit. This represents the maximum degree of supersaturation that can be obtained before precipitation takes place. The rate of nucleation then becomes negligible and crystal growth occurs. The rate of crystal growth is dependent on reagent concentration, reaction temperature and pH and the counterion in the reactant salt.[10]

Precipitation can be carried out: (i) sequentially, (ii) at constant pH and high supersaturation, or (iii) at constant pH and low supersaturation.[9]

In the sequential method, alkali is added to transition metal nitrate solutions, for example, and metal hydroxides are precipitated out as their solubility products are exceeded as the pH increases. Components are generally precipitated out in more than one phase using this technique, and this can be a disadvantage as the highest surface area materials are usually obtained when the solid precipitates in one phase only. Furthermore, even if a single phase is formed, it is often an intermediate insoluble salt containing anions from the starting reagent. Thus, Petrov *et al.*[11] found that addition of alkali to zinc and cobalt nitrates gave a basic cobalt/zinc hydroxynitrate species of formula

$[Zn_{1.66}Co_{3.34}(OH)_{8.82}(NO_3)_{1.26}(H_2O)_{2.23}]$. Toxic nitrogen oxides are evolved when hydroxynitrates are decomposed and oxides prepared from these precursors have been found to be poor sulfur absorbents.[12]

Precipitation at constant pH and high supersaturation occurs when high concentrations (~ 4 mol dm^{-3}) of metal nitrate solutions are added rapidly to a base. Many small metal nuclei are formed under these conditions so that the rate of nucleation is much higher than that of crystal growth. This results in the eventual formation of a large number of small particles whose composition may not be uniform. The precipitate is generally amorphous[13] and the solid grows without 'knowing' what its final structure will be. However, the ions or molecules in simple compounds will rapidly reorganise themselves in order to form a regular lattice since this is favourable energetically. This is much more difficult for molecules in complex structures such as silicates, and in these cases an amorphous gel structure is formed.

Precipitation at constant pH and low supersaturation is carried out by adding dilute solutions of metal salts to a base such as sodium hydroxide or carbonate in a controlled manner such that the pH is maintained constant throughout the precipitation reaction. The rate of addition of the two solutions is controlled using peristaltic pumps, and reactions are carried out at a preset temperature. This technique is generally referred to as coprecipitation. A typical arrangement for a coprecipitation reaction is shown in Figure 5.3.

Under conditions of low supersaturation a small number of nucleation sites are generated. This means that there will be little material in solution and growth will proceed slowly, allowing the precipitated particles to adopt well

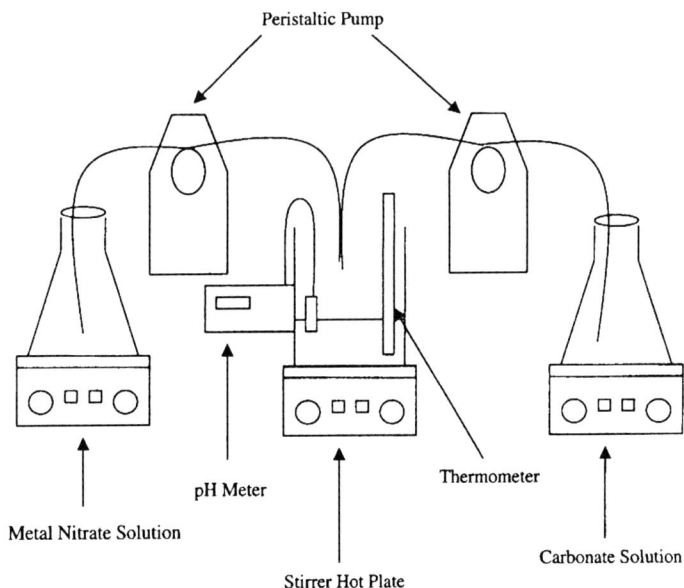

Figure 5.3 *Schematic diagram of the coprecipitation technique.*

defined shapes related to the crystal structure. Careful optimisation of each of the preparation parameters described above can be used to give an intimate mixture of metals in the same structure, and, by appropriate selection of the decomposition temperature, the metals can be retained in close contact in the same structure on calcination. These materials generally have a very high surface area.

The pH at which coprecipitation at low supersaturation is carried out is critical in determining the precipitated phases formed. This is exemplified in the preparation of layered Co/Zn basic carbonates with the hydrozincite structure. (The formula for hydrozincite is $[Zn_5(CO_3)_2(OH)_6]$ and its structure is described later in this chapter.) The hydrozincite structure was formed when using mixed Co/Zn nitrates with atomic ratios of 10/90, 20/80 and 30/70 and precipitating at a pH of 7. It was necessary to carry out the coprecipitation reaction at a pH higher than or equal to that at which the more soluble precipitating species drops out of solution. It was proposed that the precipitate would initially be predominantly $Zn(OH)_2$ since it has a lower solubility product than either the $Co(OH)_2$ or either of the metal carbonates. The carbonate or hydroxycarbonate could then be formed by an anion exchange mechanism.[14] Some entrainment of cobalt particles would then occur as the solution became depleted in zinc. Finally hydrozincite could be formed through anion exchange of the hydroxide for carbonate.

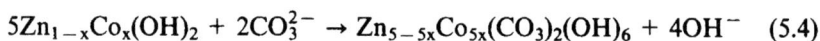

$$5Zn_{1-x}Co_x(OH)_2 + 2CO_3^{2-} \rightarrow Zn_{5-5x}Co_{5x}(CO_3)_2(OH)_6 + 4OH^- \quad (5.4)$$

At pH greater than 12 the zincate anion $[ZnO_2]^{2-}$ is formed. The composition of the precursor is also dependent on the ratios of the metal ions. Porta *et al.*[15] prepared Co/Cu hydroxycarbonates by coprecipitation at pH 8 using sodium hydrogen carbonate as the base. Precursors with a Co/Cu ratio of $< 33/67$ formed a cobalt containing malachite, $[Cu_{2-x}Co_xCO_3(OH)_2]$, whereas copper-containing spherocobaltite, $[Co_{0.85}Cu_{0.15}CO_3]$, was formed for a Co/Cu ratio of 85/15 and pure spherocobaltite, $CoCO_3$, was formed for a Co/Cu ratio of 100/0.

When the precipitation reaction is complete, the precipitate is usually aged prior to filtration and drying by stirring the precipitate in the mother liquor, *i.e.* the liquid from which the precipitate has formed. The effective precipitate surface area can change during ageing. The particles undergo agglomeration, recrystallisation and surface modifications as they attain surface equilibrium. The morphology and crystallinity of phases often show a strong dependence on the length of time for which the precipitate was aged. Thus, in the preparation of Cu supported on ZnO catalysts used in the synthesis of methanol, a zincian malachite $[Cu_{2-y}Zn_yCO_3(OH)_2]$ (Cu/Zn 85/15) and aurichalcite $[Cu_{5-x}Zn_x(CO_3)_2(OH)_6]$ were formed by coprecipitation using mixed metal nitrates with a Cu/Zn molar ratio of 2:1 at 60 °C and pH 7. On ageing in the mother liquor at 60 °C, the aurichalcite was lost and a more finely divided copper-enriched (Cu/Zn 2:1) malachite phase was formed.[16]

Precursors are of interest in their own right as sorbents, but frequently they are decomposed to their oxides before use. This step should be carried out at the minimum temperature required to form the oxide in order to achieve a high surface area. At higher temperatures there is an increased interaction between the mixed metal oxides and this can result in a loss in surface area and the formation of ordered spinels with reduced absorption capacity.

3 Layered Structures

Many of the precipitated phases used as sorbents for sulfur form layered structures and one of the most important of these is the hydrotalcite structure. The parent hydrotalcite structure is the natural mineral magnesium aluminium hydrotalcite which has the formula $Mg_6Al_2(OH)_{16}(CO_3) \cdot 4H_2O$. The structure can be considered to be derived from that of the mineral brucite, $Mg(OH)_2$. The brucite structure consists of neutral sheets of edge-sharing magnesium hydroxide octahedra.[17] Some of the Mg^{2+} ions are then replaced by Al^{3+} ions in order to generate the hydrotalcite structure. This means that the $Mg^{2+}-Al^{3+}-$ OH sheet gains one positive charge for each Mg^{2+} replaced by Al^{3+}. An equivalent number of anions must then be placed between the positive sheets to balance the charge. Water molecules are also found between the cationically charged sheets. These layered structures are known as hydrotalcites or, on account of the anions distributed between the layers, anionic clays. A representation of the hydrotalcite structure is shown in Figure 5.4.

The hydrotalcites can be contrasted with the naturally occurring cationic clays such as montmorillonite. The clays are formed from sheets of negatively

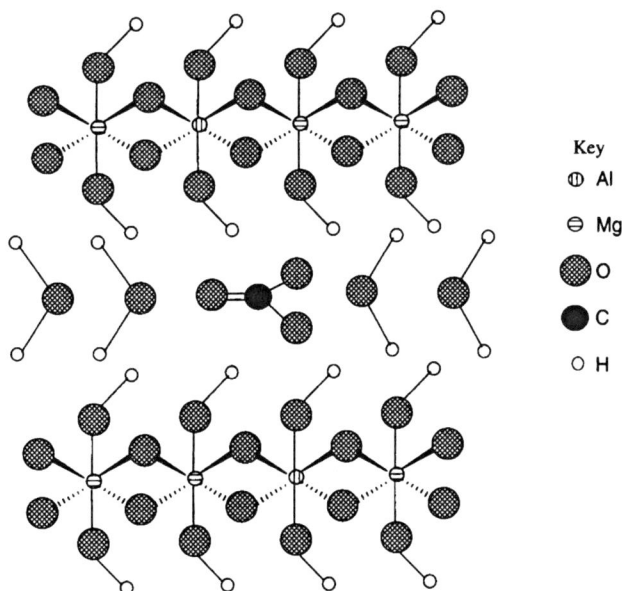

Key

① Al

⊖ Mg

⬤ O

⬤ C

○ H

Figure 5.4 *Schematic representation of the hydrotalcite structure.*

charged $[SiO_4]^{4-}$ tetrahedra and octahedral $[Al_2(OH)_6]$ units. In mont-morillonite, the octahedral Al^{3+} are partially substituted by Mg^{2+} and the charge is balanced by hydrated cations between the layers. The tetrahedral and octahedral sheets are arranged alternately in a 2:1 ratio. The unit cell contains 20 oxygens and four hydroxyl groups and there are eight tetrahedral and six octahedral sites. Montmorillonite has the general formula $M_{x/n}^{n+}yH_2O[Al_{4.0-x}Mg_x](Si_{8.0})O_{20}(OH)_4$. The intercalated cations occupy two thirds of the octahedral sites and they are generally alkaline earth metal cations such as Ca^{2+} or alkali metal cations such as Na^+.[18,19]

A comparison of the cationic and anionic clays is illustrated schematically in Figure 5.5. The hydrotalcite structure is found in a number of minerals and synthetic materials. Another example of a naturally occurring hydrotalcite is the mineral takovite $Ni_6Al_2(OH)_6(CO_3) \cdot 4H_2O$. Synthetic materials with the hydro-talcite structure generally have a combination of M^{2+} and M^{3+} ions such as $Zn_6Cr_2(OH)_{16}(CO_3) \cdot 4H_2O$ and $Co_6Al_2(OH)_{16}(CO_3) \cdot 4H_2O$. They can be pre-pared by coprecipitation at constant pH (pH 7–10 generally used depending on the metal salts) from soluble salts of the metals and bases as described earlier. The precipitate is usually aged in the mother liquor before filtering, washing and drying. Pure hydrotalcites can only be synthesised by careful control of factors such as the M^{2+}/M^{3+} ions and their ratio, the pH at which the precipitation is carried out and the temperature at which the precipitation and ageing stages of the reaction are carried out.[17] The criteria for the choice of metal ions is that their ionic radii have to be similar to that of Mg^{2+} in order that they can be accommodated in the brucite-like layered structures. Thus, for example, the divalent ions Mg^{2+}, Ni^{2+}, Co^{2+}, Zn^{2+}, Fe^{2+} and Mn^{2+} can form hydrotalcite structures with the trivalent ions Al^{3+}, Fe^{3+} and Cr^{3+}.[9] On the other hand, Cu^{2+} will only form a hydrotalcite if it is combined with another divalent ion with a similar ionic radius as well as an M^{3+} ion, *e.g.* as in Cu-Co-Cr-CO_3. This is due to the Jahn–Teller distortion found in Cu^{2+} which makes it energetically favourable for the Cu^{2+} to exist in a discrete distorted octahedral environment rather than be accommodated in a regular octahedral environment in the brucite sheets when the ratio of Cu^{2+}/M^{2+} is greater than one.[9] The importance of the M^{2+}/M^{3+} ratio in synthesising a pure hydrotalcite is illustrated in the case of materials formed from Mg/Al and Ni/Al salts and base. If the M^{2+}/M^{3+} ratio is one, aluminium hydroxide is formed in addition to the respective hydrotalcites. If the ratio is 2–3 then the hydrotalcite is the only phase formed. However, if the ratio is four, Mg-Al hydrotalcite is formed in the case of the Mg/Al system, whereas $Ni(OH)_2$ is formed as well as the Ni-Al hydrotalcite in the Ni/Al system.[17]

Hydrotalcites can be synthesised that contain anions other than carbonate between the layers. These include inorganic ions such as Cl^- and OH^-, heteropolyacids such as $[PMo_{12}O_{40}]^{3-}$ and organic acids such as malonic acid.[9] It is difficult to prepare hydrotalcites containing anions other than carbonate directly as the carbonate anion is very stable. The anion replacement can be accomplished however by anion exchange. Thus 0.01 mol dm^{-3} solutions of HCl, HNO_3, and H_2SO_4 have been reacted with Mg-Al and Ni-Al

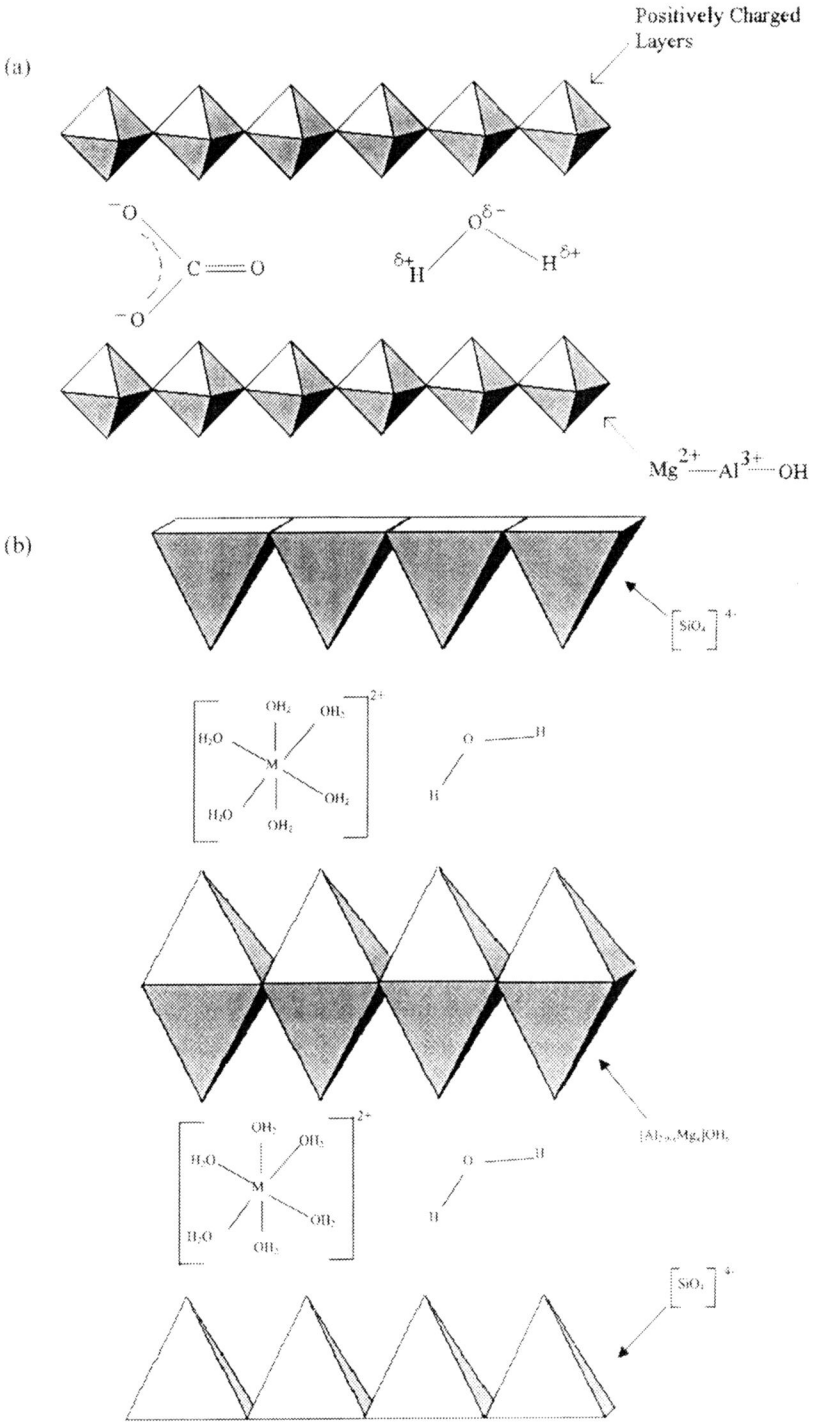

Figure 5.5 *Schematic diagram of* (a) *the anionic and* (b) *the cationic clay structures.*

hydrotalcites to form their chloride, nitrate or sulfate analogues by anion exchange.[20] The hydroxide form of the hydrotalcite can also be generated by exchange with aqueous KOH. Alternatively, the hydrotalcite can be calcined to its oxide and then on introducing a solution of the desired anion the hydrotalcite structure can be reformed. This ability of the clay structure to reform is known as the 'memory effect' and the success of the technique is dependent on factors such as the calcination time and temperature and the exchange ability of the new anion.

There is a number of layered materials which are structurally related to hydrotalcites and which are important as sorbents. Two which have been used as sorbents for H_2S are hydrozincite and feroxyhyte,[3,4] and their structures are detailed below. Hydrozincite $[Zn_5(CO_3)_2(OH)_6]$ has a layered structure and is comprised of sheets of octahedrally coordinated zinc with the formula $[Zn_3(OH)_6O_2]^{4-}$. Tetrahedrally coordinated zinc and carbonate atoms are positioned between the sheets and they balance the charge. One of the oxygen atoms from each carbonate forms part of the sheet structure. Overall, the octahedral/tetrahedral zinc ratio is 3:2 in the hydrozincite structure.[21] Feroxyhyte (δ'-FeOOH) also has a layered structure.[22] It is comprised of sheets of Fe^{3+}-anion octahedra with the O^{2-} and OH^- anions distributed in the sheets in a disordered array. The anions balance the charge within the sheets so that, unlike hydrotalcites, there are no ions present between the layers.

4   Characterisation Techniques for Sorbents

A number of techniques is available for the routine characterisation of solids and will be discussed here with reference to sorbents used in the clean-up of sulfur in industrial feedstocks such as the hydrotalcite and similar layered type structures discussed above and their products of decomposition.

X-ray Powder Diffraction (XRD)

XRD is used to provide information on the crystalline compounds or phases present in a sample. X-rays are generated by impinging a metal target such as copper with high energy electrons in a vacuum. Figure 5.6 shows a schematic diagram of the inner energy levels for copper and the generation of K_α X-rays. Figure 5.6(a) shows the original electron occupancy of the Cu atom when a high energy electron beam impinges on it. The nomenclature 'K,L,M' refers to the first, second and third principal quantum numbers and is the notation used for X-rays. The incident electrons have sufficient energy to eject electrons from the 1s shells of the copper atoms [Figure 5.6(b)]. Electrons from outer orbitals then drop down to fill the 1s levels in the copper atoms and energy is released in the form of X-rays [Figure 5.6(c)].

The transitions are quantised and a fixed amount of energy is released for each transition to give a spectrum of characteristic X-rays. An electron falling from a 2p to a 1s level is known as a K_α transition and the energy released gives rise to a characteristic X-ray that has an energy equivalent to a wavelength of

High Energy Electron Beam

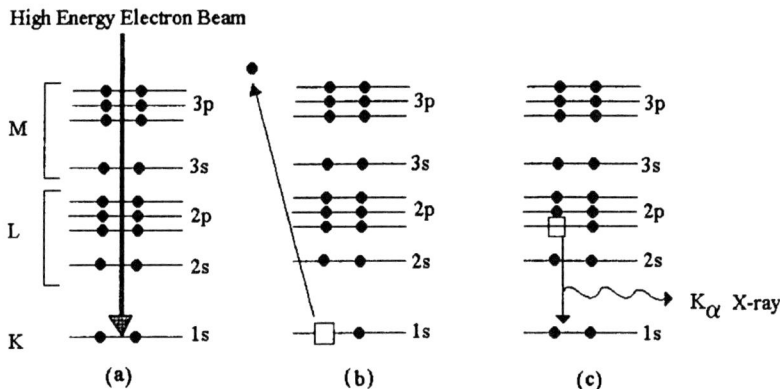

Figure 5.6 *Schematic diagram showing the generation of K_α X-rays in a copper target.*

1.5418 Å in the case of copper. This is the energy that is used in diffraction experiments as it has the greatest intensity. The K_β transition arises from an electron dropping from a 3p to a vacated 1s level and in copper it has an energy corresponding to a wavelength of 1.3922 Å. These characteristic X-rays lie on top of a broad signal of continuous radiation known as white radiation or Bremsstrahlung which arises when electrons collide with nuclei and lose energy as electromagnetic radiation. In the limiting case, electrons give up all their energy in a single encounter with a nucleus so that there is a minimum wavelength for a given target material below which the white radiation is not observed. For X-ray diffraction experiments a nickel filter is used for the Cu K_α radiation in order to absorb the K_β and most of the white radiation so that the X-ray beam is essentially monochromatic. Absorption increases with wavelength for a given element but discontinuities are observed which correspond to the ionisation energy of the various electronic levels of the atom and at these points a pronounced drop in absorption is observed. These discontinuities are known as absorption edges, and the one at highest energy which corresponds to the ionisation of a 1s electron is known as the K edge. Nickel is effective as a filter since its K edge at $\lambda = 1.488$ Å lies between that for the Cu K_α and Cu K_β transitions. The Cu K_β radiation, which is higher in energy (E) than the K_α radiation ($E \propto 1/\lambda$), will ionise a Ni 1s electron which is thus absorbed by the filter, but the Cu K_α has a lower energy than the nickel and thus passes straight through. The white radiation has a range of energies many of which will be removed by the filter. The emerging beam will thus be comprised of Cu K_α and a small amount of white radiation.

Diffraction is the interference that results from having an object in the path of waves. The interference gives a pattern of varying intensity depending whether interference is constructive (waves add) or destructive (waves cancel each other out) and this is known as the diffraction pattern. A condition for the diffraction to occur is that the dimensions of the diffracting object have to be similar to that of the wavelength of the radiation. Thus, sound waves, which have a wavelength of *ca.* 1 m, are diffracted by macroscopic objects 1 m wide, light waves with a

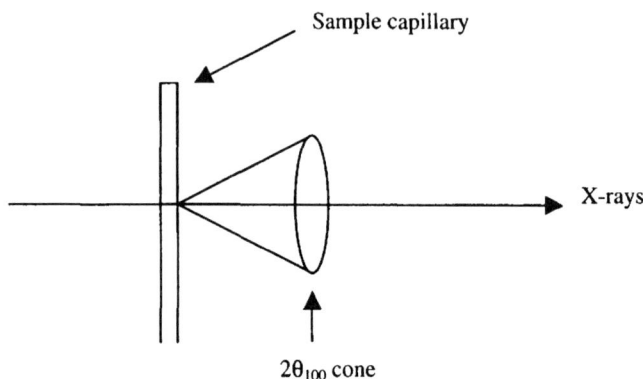

Sample capillary

X-rays

$2\theta_{100}$ cone

Figure 5.7 *Schematic diagram of a cone generated from a set of lattice planes.*

wavelength *ca.* 500 nm will be diffracted by narrow slits approximately 100 nm across and X-rays which have wavelengths comparable to the spacings of atoms in crystals (*ca.* 100×10^{-3} nm) are diffracted by the regular arrays of atoms or ions in crystalline solids, *i.e.* the crystal or lattice planes. The diffracted radiation from a given set of lattice planes will form a cone (Figure 5.7).

This is because there is no restriction on the orientation of these planes relative to the incident X-ray beam and, since each crystal in a powdered sample will be at a slightly different angle to the incident beam, each will give its own diffracted beam and collectively they will form a cone of diffracted radiation. Each set of lattice planes will form its own cone of radiation. The diffracted cones in successive planes will cancel each other out unless they are in phase, and to be in phase they must obey Bragg's law. This states that:

$$n\lambda = 2d\sin\theta \tag{5.4}$$

where
λ is the wavelength of the X-rays (nm),
d is the distance between the lattice planes (nm), and
θ is the angle of incidence of the X-ray beam to the lattice plane.

In practice, the powdered sample is mounted in a small glass capillary in the centre of a cylindrical camera known as the Debye–Scherrer camera. A nickel filtered Cu K_α X-ray beam is used. An X-ray beam is passed through one side of the camera and focused onto the sample which is rotated so that a large number of orientations of the crystallites are exposed to the beam. The diffracted cones can be detected using a thin strip of film which is placed around the sample in the camera. The cones form arcs on the film on either side of the undiffracted beam. The separation between pairs of corresponding arcs is measured and the *d* spacings can be calculated by substituting into the formula:[6]

$$\frac{S}{2\pi R} = \frac{4\theta}{360} \tag{5.5}$$

where
S is the separation between corresponding arcs, and
R is the radius of the camera film.

The main disadvantages of this method are that exposure times of up to 24 hours are required and closely spaced arcs are not well resolved.

An alternative form of detection is to use a diffractometer which is comprised of a moveable detector such as a Geiger–Müller tube or a scintillation detector connected to a chart recorder. The detector is moved in the circumference of a circle around the sample measuring the diffraction cones at the diffraction maxima for each of the lattice planes. The intensities of the X-rays are measured with respect to 2θ, where 2θ is the angle between the detector and the sample. The *d* spacings which represent the distance between the lattice planes for each of the peaks can then be calculated from Bragg's law, and then the phases present in the sample can be identified by comparing peak positions with those of known standards. Each of the planes is denoted by three Miller indices (see Chapter 6). Over 50 000 X-ray powder diffraction patterns are available in the data base prepared by the Joint Committee on Powder Diffraction Standards (JCPDS).[23]

XRD is used extensively in the characterisation of hydrotalcites even though the diffraction patterns are often broad and poorly defined due to low crystallinity. Furthermore, disorder in the stacking of the layered structure is often observed. This lowers the symmetry and changes the intensities of the peaks relative to that of the hydrotalcite standard.[9] The XRD pattern is shown for a Cu-Co-Al hydrotalcite in Figure 5.8.[9,13] The Miller indices associated with each of the main d spacings are shown in the figure.

The intense and well defined lines at low 2θ values correspond to the basal plane spacings, *i.e.* the separation between the layers in the hydrotalcite structure. The size of the interlayer spacing will depend on the anion size.

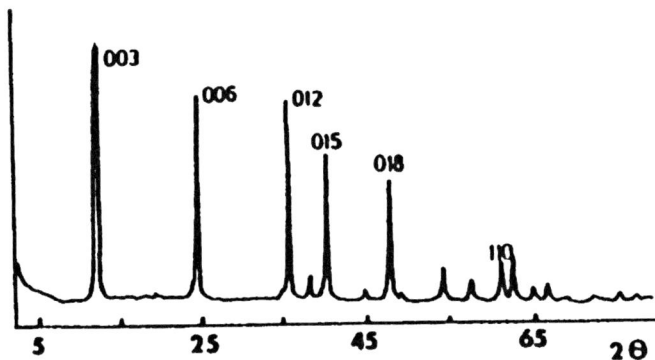

Figure 5.8 *XRD profile for a Cu-Co-Al hydrotalcite.*
(Adapted from *J. Mol. Catal.*, P. Courty, D. Durand, E. Freund and A. Sugier, 'C_1–C_6 alcohols from synthesis gas on copper cobalt catalysts', p. 246. © 1982, with permission from Elsevier Science)

Infra-red Spectroscopy

Infra-red (IR) spectroscopy is used to detect vibrations of atoms in solids by varying the frequency or wavenumber of incident radiation and measuring the amount of radiation transmitted or absorbed by the sample with respect to wavenumber. Groups of atoms in solids vibrate with frequencies of the order of 3×10^{12} to 3×10^{14} Hz and they can be excited to higher energy levels by absorbing radiation. The separations between energy levels are of the order of 10^4 J mol^{-1}.[24] The vibrations of the molecules are internal oscillations about the equilibrium bond lengths and each has a characteristic vibration frequency that is given by equation 5.6:

$$v = \frac{1}{2\pi c}\sqrt{\frac{k}{\mu}} \quad cm^{-1} \tag{5.6}$$

where
c is the velocity of light (used to convert from frequency to wavenumbers, which are the standard IR units),
μ is the reduced mass for the system, given for two masses m_1 and m_2, by equation 5.7,

$$\mu = \frac{m_1 m_2}{m_1 + m_2} \tag{5.7}$$

and k is the force constant.

The force constant arises since the compression or extension of a bond during the vibration of a molecule is analogous to the behaviour of a spring which obeys Hooke's law. Hooke's law states that within the elastic limit a strain is proportional to the stress producing it, and this can be expressed as:

$$f = -k(d - d_{eq}) \tag{5.8}$$

where
f is the force required to restore the spring from its compressed or extended position to equilibrium,
d is the extent of compression or extension,
d_{eq} is the equilibrium position of the spring, and
k is the force constant.

The vibrational energy levels of atoms are quantised. The allowed vibrational levels for a harmonic oscillator obeying Hooke's law are given by:[25]

$$E = (v + \tfrac{1}{2})hv \quad joules \tag{5.9}$$

where
v is the vibrational quantum number (v = 0, 1, 2, etc.) and
h is Planck's constant.

This can be expressed in spectroscopic units as:

$$\varepsilon = \frac{E}{hc} = (v + \tfrac{1}{2})v \quad \mathrm{cm}^{-1} \tag{5.10}$$

The Heisenberg Uncertainty Principle states that both the position and momentum of an atom cannot be determined at the same time and therefore all molecular vibrations must have some vibrational energy even at absolute zero.[25] Furthermore, real molecules do not obey Hooke's law since the bonds will break if they are compressed and extended to any great extent. They are said to undergo anharmonic oscillations and the allowed vibrational levels are given by:[25]

$$v = v_e[1 - x_e(v + \tfrac{1}{2})] \tag{5.11}$$

where

v_e is the equilibrium oscillation frequency for the anharmonic oscillator and represents very small vibrations about equilibrium, and

x_e is the anharmonicity constant and is small and positive for bond stretching vibrations so that the energy levels become more closely spaced with increasing v.

In order for a vibrational mode of a molecule to be IR active, the dipole moment must vary during the vibration. This means that centrosymmetric vibrational modes will be IR inactive and that diatomic homonuclear molecules such as dinitrogen will not have an IR spectrum. The number of fundamental vibrations, not all of which may be IR active, can be determined from the number of degrees of freedom in the molecule. A molecule with N atoms can have $3N$ degrees of freedom in three dimensional space. The molecule can undergo translational movement and this results in the loss of three degrees of freedom (one in each of the directions x, y and z as defined in cartesian coordinates). A non-linear molecule can also undergo rotation about the three axes and this removes another three degrees of freedom, leaving $3N$-6 degrees of freedom which are the internal vibrations known as the fundamental vibrations. In the case of a linear molecule, there is no rotation around the bond axis so there will be $3N$-5 fundamental vibrations.

The IR spectra of solids are usually very complex and a large number of peaks is observed, where each one can be assigned to a particular vibrational transition. However, since vibrations in which there is not a change in dipole moment are IR inactive, the number of peaks will generally be less than the number of vibrational modes. The vibrations described above are fundamental vibrations characteristic of groups of molecules within the solid for which the change in vibrational energy level $\Delta v = \pm 1$ but overtone and combination bands can also be observed. Overtone bands are much smaller in intensity than fundamentals and arise from transitions in which $\Delta v = \pm 2$ or ± 3.

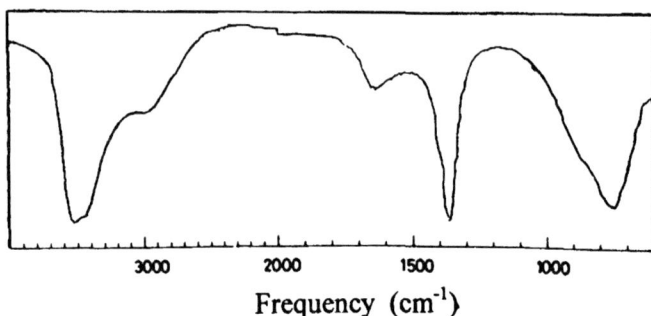

Frequency (cm⁻¹)

Figure 5.9 *IR spectrum of a Ni-Al-CO₃ hydrotalcite.*
(Reproduced from ref. 26, figure 2a, by permission of the Royal Society of
Chemistry.)

Combination bands arise from the addition of fundamental frequencies or
overtones and again they are very weak.

IR spectroscopy can be used to identify the presence of anions between the
layers of the hydrotalcite structure. It can also provide information on the
types of bonds formed and their orientations. Figure 5.9 shows the IR spectra
for Ni-Al-CO$_3$ as a typical hydrotalcite.[26]

An absorption band is observed at *ca.* 3500–3600 cm^{-1} which can be assigned
to hydrogen bonding stretching vibrations of the OH group in the hydrotalcite
layer structure.[9] Serna *et al.*[27] found that the hydrogen stretching and bending
frequencies in hydrotalcites increased as the ratio of M(II)/M(III) cations
increased from 2 to 3. He attributed this shift to a modification of the layer
spacing.

The anions lying between the layers in the hydrotalcite structure give IR
absorption bands in the region of the spectrum between 1000 and 1800 cm^{-1}.
Taking the example of the carbonate anion, which has D$_{3h}$ planar symmetry, six
($3N - 6$) bands would be expected. The vibrational modes are shown in Figure
5.10.

Modes v_3 and v_4 are degenerate,[28] there being two vibrational modes with the
same energy for each of v_3 and v_4. v_1 is IR inactive as no change in dipole

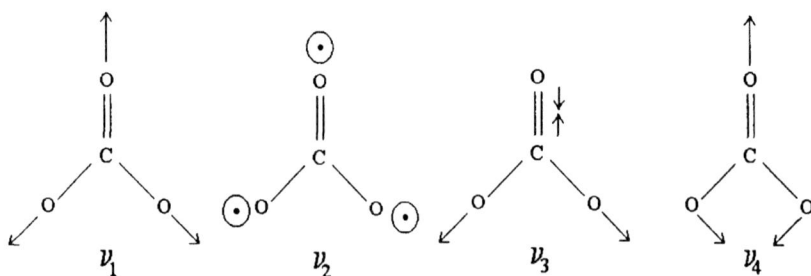

Figure 5.10 *Vibrational modes of the carbonate anion.*

moment is observed during this vibration. These three bands are also observed in the IR spectrum of the hydrotalcite in the region 1350–1380 cm^{-1} (v_3), 850–880 cm^{-1} (v_2) and 670–690 cm^{-1} (v_4).[9] In some cases, however, either an additional shoulder or peak has been observed between 1350 and 1400 cm^{-1}. This has been attributed to a lowering of the symmetry of the carbonate from D_{3h} to C_{2v} in the hydrotalcite and to the space between the layers being distorted so that the v_3 and v_4 modes are not degenerate.[9,29] The v_1 mode, which occurs at *ca.* 1060 cm^{-1}, is inactive in D_{4h} symmetry but becomes active in this lowered symmetry. Miyata[9,30] suggested that the carbonate could form monodentate and bidentate complexes between the hydrotalcite layers and that this was responsible for the lowering of the symmetry. In contrast, Serna *et al.*[9,31] observed a band at 1625 cm^{-1} and attributed this to bicarbonate ions, and interpreted the shoulder or additional band in the 1380 cm^{-1} region and the appearance of a band at 1060 cm^{-1} as being due to a distortion of the carbonate anion under vacuum conditions.

UV-vis–Near-IR Diffuse Reflectance Spectroscopy

Whereas IR spectroscopy covers the spectral range from 4000 to 300 cm^{-1} and measures changes in the vibrational energies of molecules, UV-vis–near-IR diffuse reflectance spectroscopy covers the spectral range from 50 000 to 5000 cm^{-1} and measures changes in the electronic energies of molecules. The separation between the energy levels is *ca.* 100–1000 kJ mol^{-1}.

Electronic transitions can occur in one of several ways. Firstly, an electron can be promoted from one orbital of an atom to another orbital of higher energy in the same atom such as is observed for transition metals with partially filled d orbitals. Thus, in $[Fe(H_2O)_6]^{2+}$, the iron is octahedrally coordinated by six water molecules and has the ground state electron configuration $[Ar]3d^6$. In an octahedral crystal field the five d orbitals are split into the lower energy t_{2g} set of orbitals comprised of the d_{xy}, d_{xz} and d_{yz} orbitals, and a higher energy e_g set of orbitals comprised of the $d_{x^2-y^2}$ and d_{z^2} orbitals. The e_g orbitals are higher in energy than the t_{2g} orbitals since the former point directly at the ligands whereas the latter are directed between the ligands [Figure 5.11(a)]. The energy level diagram for $[Fe(H_2O)_6]^{2+}$ is sketched in Figure 5.11b.

It can be seen that $[Fe(H_2O)_6]^{2+}$ adopts a high spin configuration in which the number of unpaired electrons is maximised. This is because the energy required to pair up electrons in orbitals (the pairing energy) is much greater than the octahedral crystal field, Δ_o, which is the separation between the t_{2g} and e_g orbitals. The promotion of an electron from the t_{2g} to e_g orbital is known as a d–d transition. These transitions are usually weak as they break the quantum mechanics selection rule. This states that in a system that has a centre of symmetry (as is the case for an octahedral complex), an electron has to move from an orbital that is even to one that is uneven (or *vice versa*) with respect to inversion through the centre of symmetry during an electronic transition. This selection rule is disobeyed since all d orbitals are even with respect to inversion through their centre of symmetry. However, transitions are observed since all

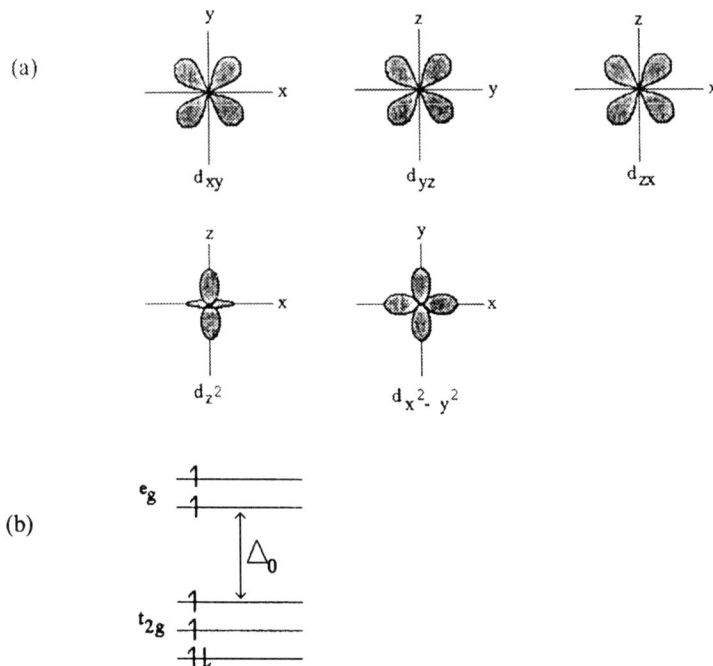

Figure 5.11 *d Orbitals in* $[Fe(H_2O)_6]^{2+}$: (a) *spatial configuration,* (b) *energy level diagram.*

molecules vibrate and these vibrations distort the complex sufficiently to partially remove the centre of symmetry. The absorption bands for tetrahedral complexes are much more intense since tetrahedral complexes do not have a centre of symmetry and so the quantum selection rule does not apply.

Electronic spectra can also arise from charge transfer interactions. These involve the transfer of an electron from: (a) a metal orbital to a ligand orbital (metal→ligand charge transfer), or (b) a ligand orbital to a metal orbital (ligand→metal charge transfer). The transfer of electrons between metal and ligands is allowed and does not violate any selection rules so the spectra tend to be much more intense than those associated with d–d transitions. The ligand to metal charge transfer transition is the most common and their spectra are usually found in the vis–near-UV region of the spectrum. They are generally found for complexes which have electronegative ligands and metals in high oxidation states, an example being iron thiocyanate [Fe(SCN)$_3$] which is deep red due to electron transfer from the thiocyanate to Fe^{3+}.[32] Metal to ligand charge transfer transitions are found for metals in low oxidation states and ligands with vacant π^* orbitals as in Mo(CO)$_6$, where an electron is transferred from the Mo t_{2g} to the CO π^* orbitals.[32] These transitions are again allowed and are generally seen in the visible part of the spectrum. A third type of charge transfer interaction is the metal–metal or intervalence charge transfer which occurs when a metal is present in two different oxidation states, as observed, for example, in Co_3O_4, which has a normal spinel structure with Co^{2+} tetrahedrally coordinated and Co^{3+} octahedrally coordinated, *i.e.* $[Co^{2+}]^{tet}[Co_2^{3+}]^{oct}O_4$. Electron transfer between Co^{2+} and Co^{3+} gives rise to a charge transfer band across the visible region.[33]

The UV-vis spectra of solids that are insoluble in common solvents such as the hydrotalcite based materials are determined using a technique known as diffuse reflectance. This technique involves indirectly measuring the absorption of the powdered solid as a function of wavelength by looking at the reflection from the solid. It is not possible to press the sample into a disc with potassium bromide (as is the case in IR) and then measure the transmission of radiation through the sample since too much scattering occurs at the shorter wavelengths of UV-vis radiation. Measurements using the diffuse reflectance technique are complicated by the presence of multiple scattering, making the spectra difficult to interpret. However, this problem has been overcome with the development of the Kubelka–Munk theory which treats absorbance and scattering separately. The spectra are determined under experimental conditions where reflection is mainly diffuse, *i.e.* it occurs in all directions independent of the angle of incidence. This can be contrasted with specular reflectance where the reflected beam travels back along its incident path. The Kubelka–Munk theory thus assumes that all the reflection is diffuse. Terms for absorbance and scattering are developed[34,35] and the final equation can be written as:

$$F(R_\infty) = \frac{(1 - R_\infty)^2}{2R_\infty} \tag{5.12}$$

where $F(R_\infty)$ is the Kubelka–Munk function and is equal to k/s, where k is $2 \times$ the fraction of radiation absorbed per unit path length and s is $2 \times$ the fraction of radiation scattered per unit path length. The factor of two in k and s arises since radiation is coming from both above and below a unit area of sample. R_∞ refers to the diffuse reflectance from a layer of infinite thickness. In practice, provided that the sample is a few millimeters in thickness it can be regarded as infinitely thick for the purpose of reflectance measurements. The particle size should be between 0.1 and 1 μm, since specular reflection increases significantly at higher particle size and causes the spectrum to be flattened, and the scattering becomes very dependent on the wavelength below 0.1 μm.[34] The reflectance is measured relative to a highly reflective standard (usually $BaSO_4$) and the reflected light is collected in a $BaSO_4$- or MgO-coated integrated sphere connected to a chart recorder via a photomultiplier detector for the UV-visible range and a lead sulfide detector for the near-IR range.[35]

Figure 5.12 shows the UV-vis–near-IR diffuse reflectance spectrum for $Co_3Zn_3Al_2(OH)_{16}CO_3 \cdot 4H_2O$ which has been shown by XRD to adopt the hydrotalcite structure.[36] The ordinate axis is R_∞^1, which is equal to $R_{\infty(sample)}/R_{\infty(standard)}$.[34]

UV-vis–near-IR diffuse reflectance can be used to assign the oxidation state and local environment of the metal cations in the hydrotalcite layered structure.

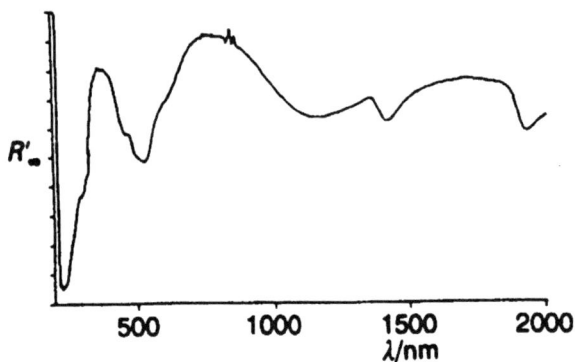

Figure 5.12   *UV-vis–near-IR diffuse reflectance spectrum for a Co-Zn-Al hydrotalcite.*
(Reproduced from ref. 36, figure 3a, by permission of the Royal Society of
Chemistry)

An absorption edge was observed at *ca.* 372 nm. This edge was also observed in
the spectrum of the hydrotalcite $Zn_6Al_2(OH)_{16}CO_3 \cdot 4H_2O$ and it can be assigned
to zinc–oxygen charge transfer. A small peak observed at *ca.* 1400 nm can be
attributed to an OH overtone.[37] Two bands were also observed in the Co/Zn
hydrotalcite at *ca.* 520 and 1150 nm which can be assigned to Co^{2+} in
octahedral coordination.[38] Finally, the three weak bands positioned at 310,
400 and 600 nm can be attributed to Co^{3+} in an octahedral environment.[39]
These assignments are in accordance with the known hydrotalcite structure
which has layers of divalent and trivalent metal cations each octahedrally
coordinated to six oxygens in two-dimensional sheets as discussed previously.

Thermal Methods

The main thermal techniques used in the characterisation of hydrotalcite clays
are thermogravimetric analysis (TGA), differential thermal analysis (DTA) and
differential scanning calorimetry (DSC).

In TGA, the mass of the sample is recorded with respect to temperature as the
sample is heated in a linear manner. Weight loss can be quantitatively assigned
to decomposition of the hydrotalcite and may occur in one or more steps.
Decomposition is considered to be complete when no further weight loss occurs
with further increase in temperature. The TGA trace for the hydrotalcite
$Co_6Al_2(OH)_{16}CO_3 \cdot 4H_2O$ is shown in Figure 5.13.

The trace shows that there was a weight loss of *ca.* 10% on heating the sample
from 25 to 217 °C which was attributed to the loss of water from between the
metal hydroxide layers. A second weight loss of *ca.* 18% was then observed
between 217 and 280 °C which could be attributed to the decomposition of the
metal hydroxide layers of the hydrotalcite structure to the metal oxides and
water together with loss of interstitial carbonate as carbon dioxide. Finally,
there was a further gradual weight loss between 280 and 1000 °C that could be
attributed to traces of carbonate and water retained on the oxides.[36]

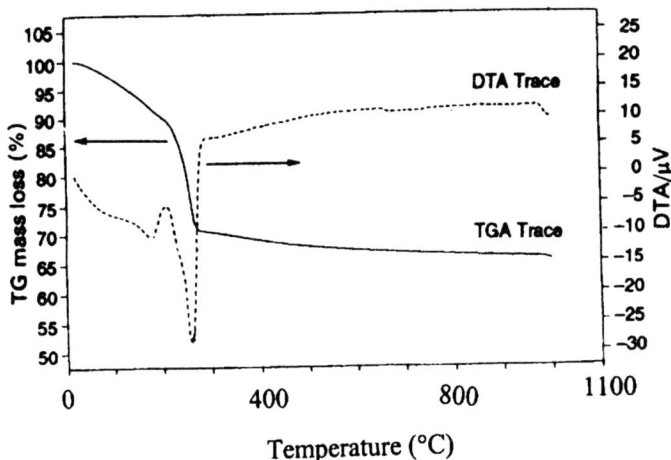

Figure 5.13 *TGA and DTA traces for a Co-Zn-Al hydrotalcite.*
(Reproduced from ref. 36, figure 4, by permission of the Royal Society of Chemistry)

In DTA, the difference in temperature of a sample compared with that of an inert reference material is measured with respect to a linear increase in temperature. A temperature difference between the sample and the reference material will arise if the sample undergoes decomposition, melts or changes crystal structure.[6] Exothermic changes will give peak maxima whereas endothermic changes will give negative peaks. The DTA trace for $Co_6Al_2(OH)_{16}CO_3 \cdot 4H_2O$ is also shown in Figure 5.13. The main features of the spectrum are the endothermic peaks at 177 °C and 261 °C which can be attributed to loss of water and carbon dioxide, respectively, as described for the TGA trace.

In DSC, the sample and an inert reference material are held at the same temperature during linear heating by inputting heat into the sample during an endothermic change or to the reference during an exothermic change. The heat input is measured and can be used to measure enthalpy changes.[6]

Electron Microscopy

Electron microscopy can provide structural information on crystalline solids to complement that obtained by XRD and it can provide details of the topography and morphology of materials. The microscope can be used at low magnification in the reflectance mode using a technique called scanning electron microscopy (SEM), or it can be used at resolutions approaching atomic dimensions in transmittance mode using transmission electron microscopy (TEM).

SEM is used over the range *ca.* 50 nm to 10 μm. Electrons are focused onto a small area of the sample which is coated with a thin layer of metal in order to prevent charge building up on the surface. Secondary electrons which are reflected by the sample and have lower energy than the incident beam are emitted from the sample. They are characteristic of the bombarded atom and can be collected and used to construct an image of the sample surface. X-rays are also emitted and since their energies are dependent on the composition of the sample they can be analysed to give an elemental map of the chemical composition of the sample. This technique is referred to as energy dispersive analysis of X-rays (EDAX) and can also be used in conjunction with a transmission electron microscope. The resolution of SEM is much poorer than that of the TEM but it gives a much greater depth of focus so that it provides valuable information on particle shape and the overall morphology of the material. It can be used to detect the shape of pores in the particles of an absorbent, for example, and thus provide an indication of whether pore diffusion contributes to its efficiency as an absorbent.

In TEM a small amount of the sample is mounted on a copper grid and covered in a thin layer of graphite, and the electron beam is passed through the sample and focused through a series of electromagnetic lenses onto a screen. The sample is highly magnified to give an image which resolves particles of the order of 1–100 nm in size. The image can be photographed for the determination of particle size and morphology, and, by adjusting the viewing screen, the diffraction pattern of a selected area of the sample can be viewed and photographed to give its phase composition. TEM has proved very useful in determining the morphology of sulfur absorbents before and after sulfiding. Figure 5.14 shows transmission electron micrographs of mixed Co/Zn oxides (a) before and (b) after passing 2% H_2S in N_2 through a bed of the sorbent until H_2S was detected in the exit stream from the bed.

The predominant phase in the oxide prior to sulfiding [Figure 5.14(a)] was ZnO ($Zn_xCo_{3-x}O_4$ was present as a minor phase) and the particles were in the form of agglomerates of small crystallites.[14] After sulfiding [Figure 5.14(b)], the particles became indistinct and were surrounded by 'sheet-like' material. Energy dispersive analysis of X-rays (EDAX) studies showed that the sheets contained cobalt, zinc and sulfur.[3]

Surface Area Analysis

The absorption capacity of sulfur absorbents is strongly dependent on the available surface area of the material, so the determination of the total available surface area of a potential sorbent is important in its characterisation. The surface area of a solid can be determined by adsorbing dinitrogen onto the solid which has been evacuated at *ca.* 10^{-4} torr to remove adsorbed gases. A plot of the amount of gas adsorbed as a function of the partial pressure of dinitrogen at constant temperature will give an adsorption isotherm from which the surface area of the solid can be calculated (Figure 5.15).

Point B is the point on the isotherm where monolayer coverage is complete. At higher pressures, multilayers of dinitrogen are adsorbed on the surface by physical adsorption and the curve then rises again sharply where vapour condenses as a liquid on the surface of the solid. If the solid contains many fine pores this curve flattens out as P/P^0 tends to one due to liquid condensing in the

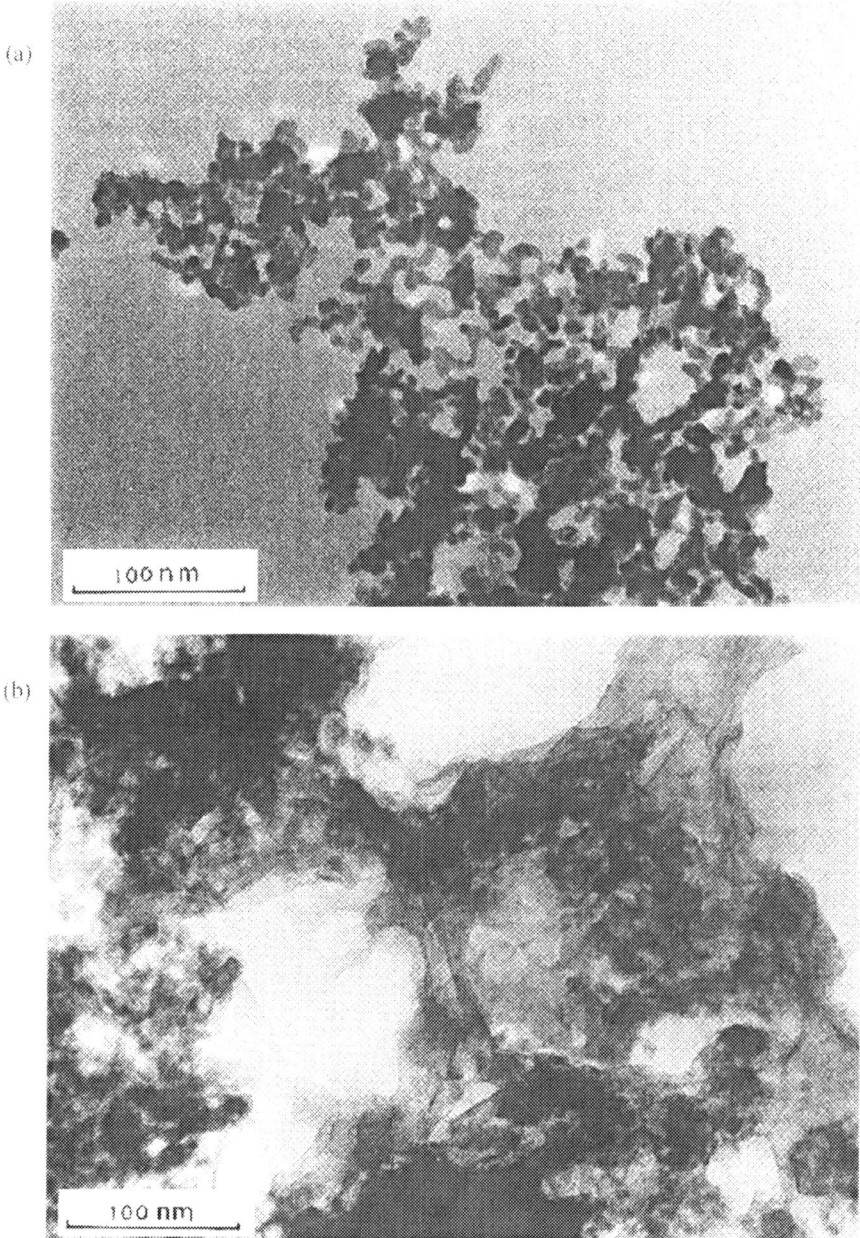

Figure 5.14 *Transmission electron micrographs of mixed Co/Zn oxides* (a) *before and* (b) *after sulfidation.*

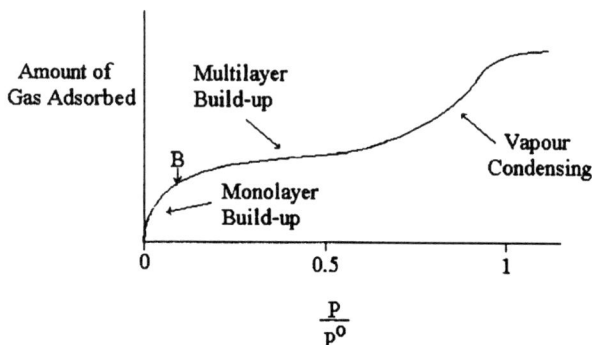

Figure 5.15 *Adsorption isotherm for a typical solid.*

pores. The theory that is used to model the adsorption process was developed by Brunauer, Emmett and Teller and is described by the BET equation.[40,41] The equation is an extension of the Langmuir theory described in Chapter 7, which describes the adsorption isotherm for chemisorption where only monolayer coverage occurs. In deriving the BET theory, the rates of evaporation and condensation of dinitrogen on both the first or monolayer and succeeding multilayers are determined and the adsorption of the monolayer of dinitrogen is assumed to be equal to the heat of adsorption of dinitrogen on the solid. The adsorption of subsequent layers of dinitrogen on the surface is considered to be equal to the heat of vaporisation of dinitrogen on the solid. It is assumed that at equilibrium, the rate of disappearance of any one layer is equal to its rate of formation, and that adsorption and evaporation only occur on exposed surfaces. The rates of adsorption and desorption are determined for each of the adsorbed layers to determine the volume of adsorbed gas with respect to pressure and the volume occupied by a monolayer. This can then be expressed in terms of the BET equation which can be written as:

$$\frac{P}{V(P^0 - P)} = \frac{1}{V_m c} + \frac{(c-1)}{V_m c} \times \frac{P}{P^0} \qquad (5.13)$$

where
V is the volume of gas adsorbed at pressure P,
P^0 is the saturated pressure of adsorbing gas at a particular temperature,
$c = \exp[(H_a - H_L)/RT]$
   where H_a is the enthalpy of adsorption of the gas on the solid and H_L is the heat produced on forming the second and higher layers, and
V_m is the volume of gas adsorbed to form a monolayer on the surface, known as the monolayer volume.

Experimentally, the surface area of the solid can be determined from a plot of $P/V(P^0 - P)$ against P/P^0 (Figure 5.16). This should give a linear plot where the intercept $I = 1/V_m c$ and the slope $S = (c-1)/V_m c$.

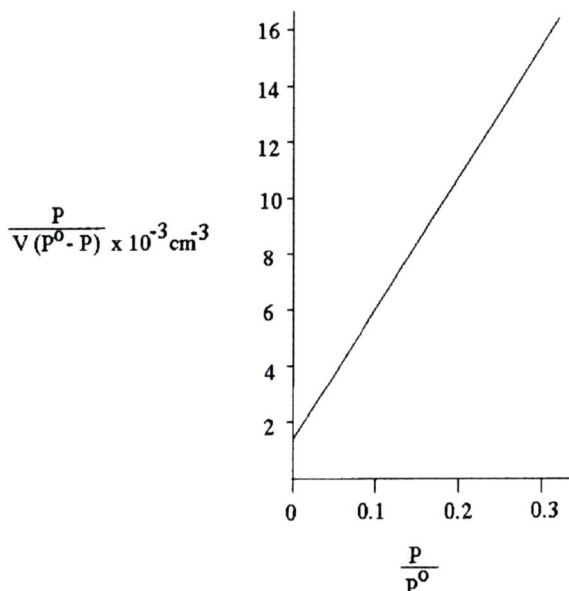

Figure 5.16 *Linear BET equation plot for determining the surface area of a solid.*

Thus $V_m = 1/(S+I)$ and this is used to determine the number of molecules of gas on the surface of the solid, *i.e.*:

$$\text{number of molecules} = V_m N_A / V_R \qquad (5.14)$$

where

V_m is the monolayer volume at standard temperature and pressure (STP) (*i.e.* $0\,°C$),

N_A is Avogadro's constant ($6.02 \times 10^{23}\,mol^{-1}$), and

V_R is the molar volume of an ideal gas ($22\,400\,cm^3$).

The surface area is then given by the number of molecules multiplied by the area per molecule. In the case of dinitrogen, the cross-sectional area when it is packed on the surface is $0.162\,nm^2$. The linear BET plot is only linear for porous solids up to P/P^0 values of *ca.* 0.3, as liquid starts to condense in the pores of solids at higher pressures. The condensation of liquid in pores can be used to determine a pore size distribution. The adsorption isotherm is determined by measuring the amount of gas adsorbed as a function of increasing pressure up to $P/P^0 = 1$. The desorption isotherm is then determined from measuring the amount of gas desorbed as a function of decreasing pressure. It is generally found that the isotherm has a hysteresis loop, *i.e.* the pressure at which liquid evaporates out of a pore will be lower than the pressure required to condense the liquid in the pore (Figure 5.17).

This can be explained by considering the Kelvin equation for the effect of

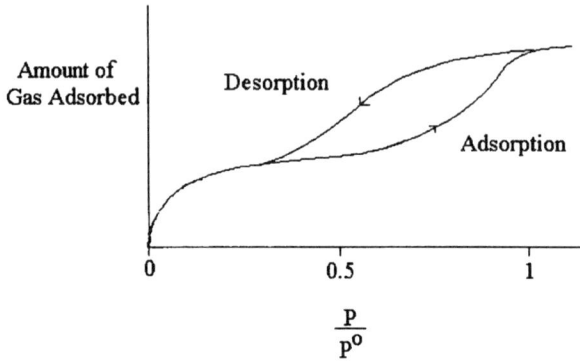

Figure 5.17 *Adsorption/desorption isotherm showing hysteresis loop.*

curvature on vapour pressure at the liquid gas interface.[41] This can be expressed as:

$$\ln \frac{P}{P^0} = -\frac{2V\gamma}{rRT}\cos\theta \qquad (5.15)$$

where
r is the radius of the pore,
V is the molar volume of the liquid,
γ is the surface tension of the liquid, and
θ is the contact angle between the liquid and the pore wall.

During adsorption, the pore will fill up with condensing vapour so that liquid will be advancing over fresh pore surface and the contact angle will be typical of that observed for liquid on a solid, *i.e.* 50–60°. The filled pore will form a concave liquid–vapour interface (Figure 5.18).

The contact angle recedes as liquid is evaporated on desorption and can be taken to be zero. The pore radius r for which vapour can condense as a liquid can be calculated for a given P/P^0 from the Kelvin equation using the data

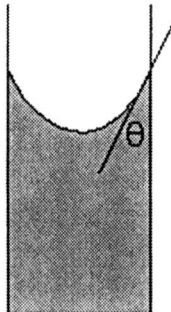

Figure 5.18 *Liquid–vapour interface for a filled pore.*

(a)

(b)

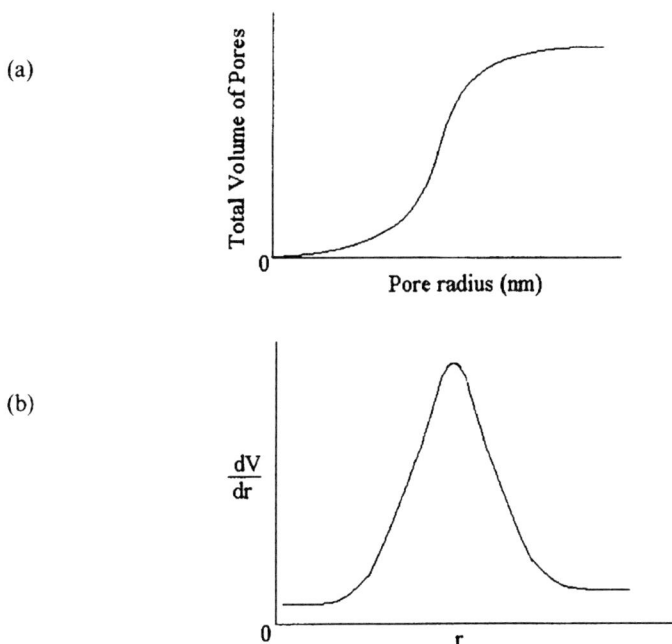

Figure 5.19 *Determination of pore size distribution:* (a) *plot of total pore volume against pore radius,* (b) *pore size distribution.*

obtained in the desorption isotherm. Vapour will condense in any pore of size r or smaller so that a cumulative plot of total pore volume against radius will be obtained [Figure 5.19(a)] and this can be differentiated to give a pore size distribution [Figure 5.19(b)]. This method is most suitable for pores up to *ca.* 30 nm as P/P^0 is too close to one at larger pore sizes.

6 References

1 P.J.H. Carnell and P.E. Starkey, *Chem. Eng.*, 1984, **408**, 30.
2 H.H. Kung, Studies in Surface Science and Catalysis, volume 45, 'Transition metal oxides: surface chemistry and catalysis', Elsevier, New York, 1991, Chapter 8.
3 T. Baird, K.C. Campbell, P.J. Holliman, R.W. Hoyle, M. Huxam, D. Stirling, B.P. Williams and M. Morris, *J. Mater. Chem.*, 1999, **9**(2), 599.
4 T. Baird, K.C. Campbell, P.J. Holliman, R.W. Hoyle, D. Stirling and B.P. Williams, *J. Chem. Soc., Faraday Trans.*, 1996, **92**(3), 445.
5 P.K. Gour, S.N. Upadhyay, S. Pande, R. Chatterjee, N.B. Bhattacharyya and S.P. Sen, Proc. Symp. Sci. Catal. Appl. Ind., Sindri, India, 1979.
6 A.R. West, 'Solid State Chemistry and its Applications', Wiley, Chichester, 1989.
7 E. Moore and L. Smart, 'Solid State Chemistry: an Introduction', Chapman and Hall, London, 1992.
8 P.J.H. Carnell, Chapter 4 in 'Catalyst Handbook', ed. M.V. Twigg, Wolfe Publishing Limited, Frome, England, 1989.
9 F. Cavani, F. Trifiro and A. Vaccari, *Catal. Today*, 1991, **11**, 173.

10 A.G. Walton, Chemical Analysis Monograph, vol. 23, 'The Formation and Proper-
    ties of Precipitates', Wiley Interscience, London, 1967.
11 K. Petrov, L. Markov, R. Ioncheva and P. Rachev, *J. Mater. Sci.*, 1988, **23**, 181.
12 R.W. Hoyle, PhD Thesis, Glasgow, 1995.
13 P. Courty, D. Durand, E. Freund and A. Sugier (IFP), *J. Mol. Catal.*, 1982, **17**, 241.
14 T. Baird, K.C. Campbell, P.J. Holliman, R.W. Hoyle, D. Stirling, B.P. Williams and
    M. Morris, *J. Mater. Chem.*, 1997, 7(2), 319.
15 P. Porta, R. Dragone, G. Fierro, M. Inversi, M. Lo Jacano and G. Moretti, *J. Mater.
    Chem.*, 1991, 1(4), 531.
16 D. Waller, D. Stirling, F.S. Stone and M.S. Spencer, *Faraday Discuss. Chem. Soc.*,
    1989, **87**, 107.
17 W.T. Reichle, *Solid State Ionics*, 1986, **22**, 135.
18 T.J. Pinnavaia, *Science*, 1983, 220(4595), 365.
19 N.N. Greenwood and A. Earnshaw, 'Chemistry of the Elements', Pergamon Press,
    Oxford, 1990, Chapter 9.
20 D.L. Bish, *Bull. Mineral*, 1980, **103**, 170.
21 S. Ghose, *Acta Cryst.*, 1964, **17**, 1051.
22 G.A. Waychimas, in 'Reviews in Mineralogy', ed. D.H. Lindsley, Mineralogical
    Society of America, Washington, 1991, vol. 25, pp. 33–38.
23 M.T. Weller, 'Inorganic Materials Chemistry', Oxford Chemistry Primers, Oxford
    Science Publications, Oxford University Press, Oxford, 1994.
24 C.N. Banwell, 'Fundamentals of Molecular Spectroscopy', 2nd edn, McGraw Hill
    Book Company, Maidenhead, 1972, Chapter 3, p. 65.
25 E.A.V. Ebsworth, D.W.H. Rankin and S. Cradock, 'Structural Methods in Inorganic
    Chemistry', Blackwell Scientific Publications, Oxford, 1987, Chapter 5, p. 164.
26 E.C. Kruissink, L.L. van Reijen and J.R.H. Ross, *J. Chem. Soc., Faraday Trans. 1*,
    1981, **77**, 649.
27 M.J.H. Hernandez-Moreno, M.A. Ulibarri, J.L. Rendon and C.J. Serna, *Phys.
    Chem. Minerals*, 1985, **12**, 34.
28 K. Nakamoto, 'Infrared and Raman Spectra of Inorganic and Coordination
    Compounds. Part A. Theory and Applications in Inorganic Chemistry', 5th edn,
    John Wiley and Sons, 1997, p. 180.
29 D.L. Bish and G.W. Brindley, *Amer. Min.*, 1977, **62**, 458.
30 S. Miyata, *Clays Clay Minerals*, 1975, **23**, 369.
31 C.J. Serna, J.L. White and S.L. Hem, *Clays Clay Minerals*, 1977, **25**, 384.
32 F.A. Cotton, G. Wilkinson and P.L. Gaus, 'Basic Inorganic Chemistry', 2nd Edn.,
    J. Wiley & Sons, Singapore, 1987.
33 A.P. Hagan, C.A. Otero, F.S. Stone and M.A. Trevethan, in 'Preparation of
    Catalysts', vol 2, ed. B. Delmon, P. Grange, P. Jacobs and G. Poncelet, Elsevier,
    Amsterdam, 1979.
34 F.S. Stone, in 'Surface Properties and Catalysis by Non-Metals', ed. J.P. Bonnelle, D.
    Riedel Publishing Company, 1983, p. 237.
35 K. Klier, *Catal. Rev. Sci. Eng.*, 1967, **1**, 207.
36 T. Baird, K.C. Campbell, P.J. Holliman, R.W. Hoyle, D. Stirling and B.P. Williams,
    J. Chem. Soc., Faraday Trans., 1995, 91(18), 3219.
37 D. Stirling and F.S. Stone, *Solid State Ionics*, 1993, **63–65**, 289.
38 J. Ferguson, *Prog. Inorg. Chem.*, 1970, **12**, 249.
39 D.S. McClure, *J. Chem. Phys.*, 1962, **36**, 2757.
40 S. Brunauer, P.H. Emmett and E. Teller, *J. Am Chem. Soc.*, 1938, **60**, 309.
41 D.J. Shaw, 'Introduction to Colloid and Surface Chemistry', 4th edn, Butterworth
    Heinemann, Oxford, 1998.

CHAPTER 6

Surface Energies and Interactions between Particles

1 Introduction

The surface of the precipitated solid will differ considerably from that of the bulk solid since each ion at the surface has a lower coordination than the ions in the solid as a whole. Each crystal thus has a surface energy which will affect the shape and surface of the growing crystals precipitated from solution. The surface energy is the energy that is needed to form unit surface area of a solid.[1] It is dependent on the packing and positions of ions in the surface. The particles are precipitated out in an aqueous medium and it is necessary to consider not only interactions between the particles themselves but also interactions between the particles and the surrounding medium. These interactions are mainly comprised of electrostatic, dispersive and hydration forces and each of these will be considered in turn in this chapter.

The positions of atoms or ions in a crystal can be referenced to the crystal planes, which are arbitrary planes governed by the shape and size of the unit cell. The planes may or may not coincide with layers of atoms in the unit cell, and they are separated by an interplanar spacing d. They are defined by numbers known as Miller indices.

2 Miller Indices

Miller indices have the general symbol (hkl). The determination of the (100) and (212) planes in a cubic crystal is illustrated in Figure 6.1.

In Figure 6.1, the origin is defined as 0 and the three crystal axes are labelled a, b, c. The following procedure is followed to assign the Miller indices to a plane.

(a) The plane is defined as the one that is adjacent to the plane passing through 0. The (100) and (212) planes are shaded in Figure 6.1(a) and 6.1(b) respectively.

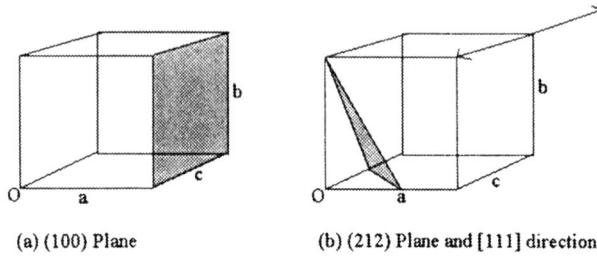

(a) (100) Plane   (b) (212) Plane and [111] direction

Figure 6.1 *Diagram showing* (a) *the (100) plane and* (b) *the (212) plane and [111] direction in a cubic crystal.*

   (b) The intersection of this plane with the three axes of the cell is determined. In the case of the (100) plane this will be 1a ∞ b ∞ c, and this is recorded as the fractional intersect 1 ∞ ∞ with the cell edges. In the case of the (212) plane this will be $\frac{1}{2}$a 1b $\frac{1}{2}$c giving the fractional intercept $\frac{1}{2}$ 1 $\frac{1}{2}$.

   (c) The reciprocals of the fractions are then taken. For Figure 6.1(a), this will give 100 and for Figure 6.1(b), this will give 212.

   (d) Where necessary, the reciprocals are multiplied through by a common factor to convert them to integers. This was not necessary in the examples given, so the Miller indices for the plane in Figure 6.1(a) are (100) and in Figure 6.1(b) (212). These indices are the Miller indices of the planes shown in the figure and all other planes that are parallel to them.

To describe a direction within a crystal, a line is drawn from the origin of the unit cell parallel to the desired direction, and the coordinates are quoted for the point at which the line emerges from the cell. Fractions are cleared to give the smallest integer. Directions are indicated in square brackets and the [111] direction is shown in Figure 6.1(b). In cubic systems, the [*hkl*] direction is perpendicular to the (*hkl*) plane.[2]

Consider now the surface energies for different planes in a sodium chloride crystal. The surface energy for the (100) surface of sodium chloride is 188×10^{-3} J/m^2, whereas it is 445×10^{-3} J/m^2 in the (110) plane.[1,3] Small precipitate particles of NaCl will adopt a compact shape which minimises the surface energy under precipitate growth conditions close to thermodynamic equilibrium, *i.e.* at low degrees of supersaturation. Cubic crystals with the (100) planes exposed will thus be formed as they have the lowest surface energy. The crystal will actually take up the shape of the unit cell since the (100) planes correspond to the atomic arrangement of the faces of the unit cell. This is often found to be the case.[1]

3 Dielectric Properties

In order to understand the extent to which particles interact in a medium, it is necessary to understand their dielectric properties. They can be defined by

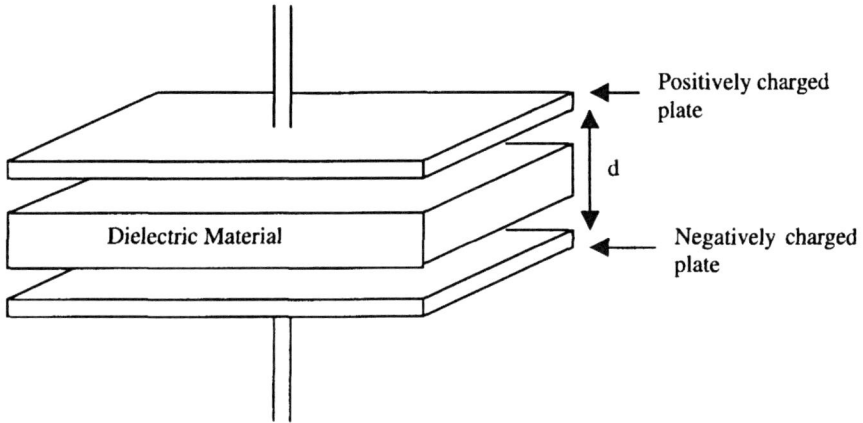

Figure 6.2 *Diagram of a parallel plate capacitor.*

considering their properties when placed in a parallel plate capacitor. This is comprised of two conducting plates to which a voltage V is applied so that one is positively charged and the other negatively charged (Figure 6.2).

The plates are separated by a distance d which is small compared to the size of the plates. When a vacuum is applied between the plates, the capacitance C_0 is given by:[2]

$$C_0 = \frac{\varepsilon_0 A}{d} \qquad (6.1)$$

where
ε_0 is the permittivity of free space, a constant with a value of 8.854×10^{-12} F m^{-1}, and
A is the area of the plates.

A quantity of charge, Q_0, will be stored on the plates, where

$$Q_0 = C_0 V \qquad (6.2)$$

If a dielectric or insulating material is placed between the plates, the capacitance and hence the charge stored will increase to C_1 and Q_1 respectively for the same applied voltage. The dielectric constant for this material ε' is then related to the increase in the capacitance by C_1/C_0. The dielectric constant ε' is a constant that is a ratio of the permittivity of the material to that of free space ε_0. The ions in the dielectric are polarised as they are spatially displaced with respect to the ion cores, and the value of the dielectric constant depends on the degree of polarisation or charge displacement that can occur in the material. The dielectric constant and polarisability varies with the applied alternating current (a.c.) frequency.[2] The dielectric constant is also important in that it relates to the solubility of ions in different solvents. Thus, many ions are soluble in water since it has a high dielectric constant.

4   Electrostatic Forces

The electrostatic forces are strong interactions that arise from the coulombic interactions between ions. Consider the interaction potential $V(r)$ between two ions. This can also be referred to as the free or available energy. The free energy for the coulombic interaction between two charges Q_1 and Q_2 separated by a distance r is then given by:[4]

$$V(r) = \frac{Q_1 Q_2 e^2}{4\pi\varepsilon_0 \varepsilon r} \tag{6.3}$$

where
ε_0 is the permittivity of free space, in $C^2 J^{-1} m^{-1}$, and
ε is the permittivity of the medium, in $C^2 J^{-1} m^{-1}$.

In the case of ions this can be expressed as:

$$V(r) = \frac{z_1 z_2 e^2}{4\pi\varepsilon_0 \varepsilon r} \tag{6.4}$$

where
z_1 and z_2 are the magnitudes and signs for the ionic charges, *e.g.* $z = 1$ for Na^+
   and $z = -1$ for Cl^-, and
e is the elementary charge (1.602×10^{-19} C).

The coulombic force F between the two ions is then given by:

$$F = -dV(r)/dr = \frac{z_1 z_2 e^2}{4\pi\varepsilon_0 \varepsilon r^2} \tag{6.5}$$

The coulombic forces hold ions such as Na^+ and Cl^- together in the lattice and the bonds between the ions are ionic. The coulombic forces thus give a measure of the binding energies between the two ions. However, in considering the ionic solid the above equations have to be modified to take into account the interaction of each ion with every other ion in the crystal. Thus in sodium chloride, for example, each Na^+ has six Cl^- ions surrounding it at a distance r (where r = radius of $Na^+ + Cl^-$), 12 next nearest neighbours at $\sqrt{2}r$, eight Cl^- at $\sqrt{3}r$, *etc.* These interaction energies can all be summed to give the total interaction energy for a pair of Na^+Cl^- ions in a lattice E_{tot}:

$$E_{tot} = \frac{-1.748\ e^2}{4\pi\varepsilon_0 r} \tag{6.6}$$

where 1.748 is referred to as the Madelung constant. The negative sign indicates that it is an attractive force, and its value varies for each crystal structure. It increases with increasing size of the ions, increasing from 1.748 for NaCl to

2.365 for $CaCl_2$ and to 4.172 for Al_2O_3.[5] The molar lattice energy can be obtained from the total interaction energy by multiplying by Avogadro's number. This gives a value of 880 kJ mol^{-1} for NaCl, which is about 15% higher than the measured lattice energy. The lattice energy of an ionic solid is the energy required to overcome the electrostatic attraction between ions in one mole of the solid and to separate them to infinity as isolated gaseous ions. The difference in these values is accounted for by the presence of repulsive forces when the ions are in contact, which will lower their binding energy. These repulsive forces are known as Born repulsive forces and are short range forces that only operate at very small interatomic distances when electron clouds of atoms overlap. If the ions are represented by hard spheres, the Born repulsive force between them is described by the repulsive potential $V(r)$:

$$V(r) = +\left(\frac{\sigma}{r}\right)^n \tag{6.7}$$

where
n is generally in the range 5 to 12,
σ is the hard sphere diameter of the ion, and
r is the sum of the radii of the ions as described previously.

$V(r)$ rapidly decreases to zero as r increases.

5 Dispersion Forces

The dispersion forces are the most important of the van der Waals forces and are always present in molecules so that they are important in the aggregation of particles in solution.

Van der Waals forces are comprised of London dispersion forces, induction and dipole–dipole forces. The dipole–dipole effects arise from the interaction between two molecules with permanent dipoles. Two dipoles in rapid thermal motion will sometimes be orientated to attract each other and sometimes they will repel each other. The overall force is generally attractive since they are closer together in attractive configurations. The induction forces arise from one molecule having a permanent dipole and inducing a dipole in another molecule. The force is attractive and depends on the polarisabilities of the molecules. Both the dipole–dipole and the induction effects give rise to a potential energy term proportional to *ca.* $1/r^6$ where r is the radius of the molecule.

The dispersion forces are always present in molecules.[4] They are long range forces that can be effective from distances ranging from interatomic spacings (*ca.* 0.2 nm) up to *ca.* 10 nm. They can best be explained by considering the interaction of two non-charged atoms such as helium. The dipole moment in one of the helium atoms is zero over a period of time since the electrons move about the nucleus and spend as much time on one side as on the other. However, at any particular instant the atom will have a finite dipole moment arising from the positions of the electrons relative to the nucleus (Figure 6.3).

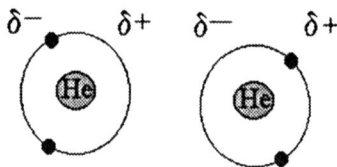

Figure 6.3 *Schematic diagram showing 'instantaneous' dipole moments arising in two helium atoms.*

This dipole will generate an electric field that will polarise an adjacent atom and thus induce a dipole of opposite charge in it. There is thus a net attractive force between these two atoms and this is known as the dispersion force. The dispersion forces are thus an induced dipole–induced dipole interaction.

Assuming that the instantaneous dipoles are oscillating with a frequency v_0, the potential U for the dispersion interaction is given by:[4]

$$U = \frac{-h v_0 \alpha^2}{(4\pi\varepsilon_0)^2 r^6} \qquad (6.8)$$

where
α is the polarisability of the molecule,
h is Planck's constant,
ε_0 is the permittivity of free space, and
r is the internuclear separation between two molecules.

The problem with the van der Waals dispersion theory described above is that it does not take into account the interactions between the molecules in a solvent medium. McLachlan[4,6] proposed a generalised theory of van der Waals forces which included terms for the induction, orientation and dispersion forces in the one equation that could be satisfactorily applied to the interactions of particles in a solvent medium. In this treatment, the excess or effective polarisabilities of the particles are measured since the polarisability will be reduced in the solvent medium due to changes in the effective dipole moment resulting from interaction of the particle with the surrounding solvent molecules.

Consider a small spherical molecule (1) of radius a_1 in a medium (2). The excess polarisability α measured at frequency v can be represented by:[4]

$$\alpha_1(v) = 4\pi\varepsilon_0\varepsilon_2'(v)\left[\frac{\varepsilon_1'(v) - \varepsilon_2'(v)}{\varepsilon_1'(v) + 2\varepsilon_2'(v)}\right]a_1^3 \qquad (6.9)$$

where ε_1' and ε_2' are the dielectric constants for the particle and medium respectively.

For a dielectric or insulating medium that has one strong absorption peak at frequency v_e which is different from that of isolated molecules in the gas phase v_I

(v_I is essentially the first ionisation energy of the molecule), the dielectric constant of the medium can be represented as:[4,7]

$$\varepsilon'(v) = 1 + \frac{(n^2 - 1)}{\left[1 - \left(\frac{v}{v_e}\right)^2\right]}$$ (6.10)

where n is the refractive index of the medium.

Assuming that the particle and the solvent phase have the same absorption frequency v_e then the total van der Waals interaction free energy for two identical particles in the aqueous medium is given by:[4]

$$w(r) = w(r)_{v=0} + w(r)_{v>0}$$ (6.11)

where
$w(r)_{v=0}$ is the zero frequency contribution to the van der Waals free energy and refers to the ionisation potential of the ion, and
$w(r)_{v>0}$ refers to the high frequency contribution to the van der Waals free energy.

The frequency contributions originate from polarisability measurements and the fact that the dielectric constants are frequency dependent. The electronic polarisability associated with atom charge displacement of an electron cloud from the nucleus will provide the dominant contribution to the dielectric constant at high frequencies where the contributions of ions or permanent dipoles or bond polarisability in covalent solids are frozen out.[8] The contributions from ionic polarisability resulting from the displacement of ions (positive and negative) in a polar solid or a permanent dipole such as in OH groups in silica contribute only at lower frequencies.[4]

The zero frequency term (term 1 in equation 6.12) measures the orientation and induction interaction energies. The second term measures the dispersion forces. Substituting in the expressions for the interaction energies:[4]

$$w(r) = -\left[\underbrace{3kT\left(\frac{\varepsilon_1'(0) - \varepsilon_2'(0)}{\varepsilon_1'(0) + 2\varepsilon_2'(0)}\right)^2}_{\text{term 1}} + \underbrace{\frac{\sqrt{3}hv_e(n_1^2 - n_2^2)^2}{4(n_1^2 + 2n_2^2)^{3/2}}}_{\text{term 2}}\right]\frac{a_1^6}{r^6}$$ (6.12)

This expression is valid for $r \gg a$,

where
r is the distance between the particles,
a is the radius of a particle,
k is the Boltzmann constant (1.381×10^{-23} JK^{-1}),

n_1 is the refractive index of the particle,
n_2 is the refractive index of the solvent, and
h is Planck's constant (6.626×10^{-34} J s).

6 Hydration Forces

The largest effect on the surface energy of a particle in solution arises from the solvent medium itself. When using polar solvents such as water there will be attractive interactions between the surface of each particle and liquid molecules at the solid–liquid interface so that structure extends several molecular diameters into the liquid (Figure 6.4).

The orientation of solvent molecules around a small particle is known as solvation (hydration when the solvent is water), and the extent of this will be governed by the geometry of the particles and how they pack together. The solvation zone will affect the dielectric constant of the solvent since the response of the solvent molecules to an electric field would be different to that of solvent molecules in pure liquid.[4] There will also be an increase in density of the solvent medium in the region where the solvation zones of particles overlap, further strengthening the solvation force. The solvation force can also be affected by specific interactions between liquid molecules. These include long range dipole polarisation and cooperative hydrogen bonding effects in the case of water. Interaction between particles may also occur, as in the case of colloidal silicas, for example, so that they cross link and form a three dimensional network and this may lead to the formation of a gel if solvent becomes trapped within the body of the solid.[9]

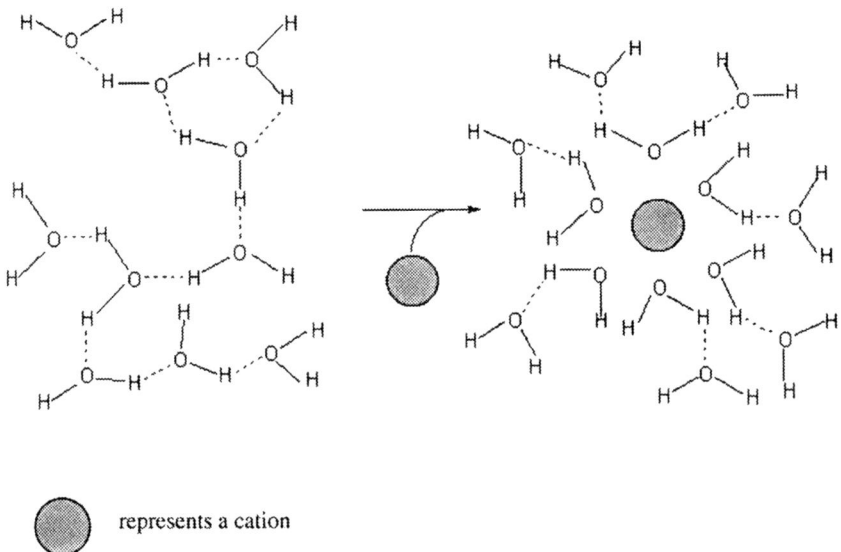

represents a cation

Figure 6.4 *Schematic diagram showing the hydration of a cation.*

7 Double Layer Effects

The growing crystal undergoes a relaxation process which lowers the surface energy. The surface becomes polarised by undergoing electronic rearrangements and anions and cations are displaced to form a charged surface. Ions of like charge are then repelled from the surface whereas ions of opposite charge are attracted towards the surface. There is some mixing of charges due to thermal motion and the combination of these effects lead to the creation of what is known as an electric double layer. Figure 6.5 shows the formation of an electric double layer for NaCl.

The double layer consists of two parts, an inner and a diffuse region.[9] The inner region is comprised of the charged surface (arbitrarily assigned a positive charge in the figure) and adsorbed ions. The surface is assumed to be flat with the charge evenly distributed on the surface. Adsorbed ions are held on to the surface by either van der Waals or coulombic forces and they are of opposite charge to that of the surface (*i.e.* they are counterions). The inner region is

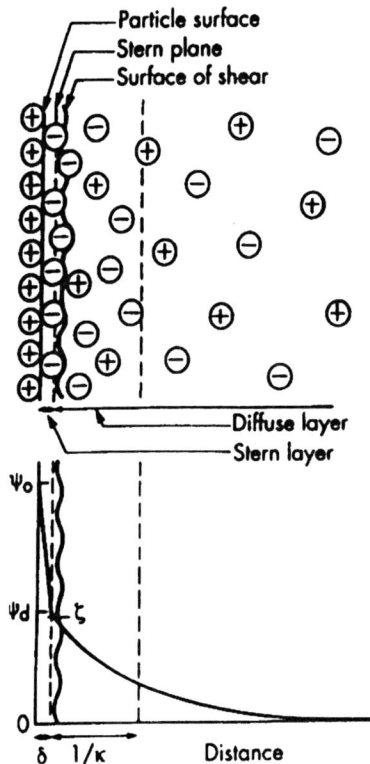

Figure 6.5 *Schematic representation of the structure of the electric double layer according to Stern's theory.*
(Reproduced from ref. 9, figure 7.2, p. 183, by permission of Butterworth Heinemann, Linacre House, Jordan Hill, Oxford, OX2 8DP. © 1992 Reed Educational and Professional Publishing Ltd)

separated from the diffuse region of the double layer by the Stern plane.[9,10] The Stern plane is found at approximately the distance of the hydrated radii of the counterions. The surface at the periphery of the counterions is known as the shear surface. The diffuse layer lies beyond the shear surface. The double layer gives rise to an electrical potential that is greatest at the surface and falls off with distance from the surface. The surface potential depends on both the ionic composition of the mother liquor surrounding the growing crystal and the surface charge density. The potential at the surface of shear is known as the zeta potential. The viscosity changes rapidly in this region due to the influence of the charged surface and the diffuse layer. Zeta potentials can be determined experimentally from electrophoresis measurements. In electrophoresis, a potential is applied across two electrodes immersed in a suspension of the particles under investigation. The particles are electrically charged and their migration towards the oppositely charged electrode is measured. Stern assumed that the adsorption of counterions in the Stern layer could be expressed using a Langmuir treatment (*i.e.* monolayer coverage and no interactions between ions; see Chapter 7).

If a molecular condenser is used as a mark for the Stern layer, the charge density at the particle surface is then given by:[9]

$$\sigma_0 = \frac{\varepsilon_s(\psi_0 - \psi_d)}{\delta} \tag{6.13}$$

where
δ is the thickness,
ε_s is the permittivity of the Stern layer,
ψ_0 is the electrical potential at the surface, and
ψ_d is the electrical potential in the Stern layer ($\psi_d < \psi_0$).

Overall, the number of positive and negative charges must balance as the double layer must be electrically neutral so that, in turn, the particle is neutral.

8 Conclusions

Overall, it is difficult to predict the behaviour of small particles in solution in terms of the electrostatic, hydration and dispersive forces and the surface relaxation processes already described. The extent to which each of these interactions is important will vary with the nature of the growing particles and the solvent medium.

For ionic solids such as sodium chloride grown from aqueous medium, the forces between the ions are uniform in all directions and electrostatic and hydration forces predominate. The crystals are hard with high melting points and will readily dissolve in solvents of high dielectric constant (polar solvents), provided that the solvation energy derived from the attractions between the ions and solvent dipoles is large enough to compensate for the loss of lattice energy when the crystals dissolve.

For covalent solids such as silica, a covalent bond network extends throughout the crystal structure to form a rigid three-dimensional framework and the solid has a high melting point and is insoluble. It is effectively a giant molecule, and the bonds within the structure are directed or orientated at well defined angles relative to one another. Here, electrostatic interactions are of less importance, unless the solid is polarised due to differences of electronegativity of the elements present and it can interact with a solvent molecule such as water. The distinction between ionic and covalent solids is often not so clear cut, though; for example, zinc sulfide has mixed ionic/covalent bonding.[2]

Molecular solids are comprised of individual discrete molecules which retain their identity in the crystal. Atoms in the molecules are held together by chemical bonds but the forces between adjacent molecules are weak van der Waals forces which result in the molecular solids being soft, volatile and low melting. An example of a molecular solid is iodine. Again, there are solids in this class that do not neatly fall into just one category. One example of this is graphite which has each carbon atom covalently bonded to three others to form two-dimensional hexagonal rings and these are stacked in flat parallel layers. Although the bonding within the rings is covalent, the bonding in the third dimension between the stacked layers is a van der Waals type of interaction.[11]

There are many other factors which may complicate the categorising of specific interactions within solids including hydrogen bonding effects either within the network of the solid or between the solid and an aqueous solvent, and interactions between solvent molecules themselves. The classification of the interactions between particles into electrostatic, dispersion and hydration forces is therefore a little esoteric and should be supported by experimental data for each case under consideration.

9 References

1 A.G. Walton, Chemical Analysis Monograph, vol. 23, 'The Formation and Properties of Precipitates', Wiley Interscience, London, 1967.
2 A.R. West, 'Solid State Chemistry and its Applications', Wiley, Chichester, 1989.
3 F. van Zeggeren and G.C. Benson, *J. Chem. Phys.*, 1957, **26**, 1077.
4 J. Israelachvili, 'Intermolecular and Surface Forces', 2nd edn, Academic Press, London, 1992.
5 D.F. Swinehart, in 'Handbook of Chemistry and Physics', 72nd edn, ed. D.R. Lide, CRC Press, Boston, 1991, **12**, 26.
6 A.D. McLachlan, *Discuss. Faraday Soc.*, 1965, **40**, 239.
7 A.R. von Hippel, 'Dielectric Materials and Applications', Wiley, New York, 1958.
8 S.R. Elliot, 'The Physics and Chemistry of Solids', John Wiley and Sons, Chichester, 1998.
9 D.J. Shaw, 'Introduction to Colloid and Surface Chemistry', 4th edn, Butterworth Heinemann, Oxford, 1998.
10 O. Stern, *Z. Elektrochem.*, 1924, **30**, 508.
11 A.K. Galwey, 'Chemistry of Solids', Chapman and Hall, London, 1967.

CHAPTER 7

Determination of the Sulfur Sorption Capacity of Solid Sorbents

1 The Sorption Processes

The transfer of a gas molecule to a solid surface involves several transport processes. The molecule must first of all move from the gas stream to the surface of the particle or pellet. It then has to diffuse through the pores of the solid to its internal surface where it is adsorbed. The adsorbed molecule is referred to as the adsorbate.

The adsorption process generally involves chemisorption in which the gas molecules form bonds to the solid and become attached to it. The maximum surface coverage is one monolayer of molecules. The adsorption process is very exothermic, with the enthalpy of adsorption of H_2S on zinc oxide, for example, being of the order of -120 kJ mol^{-1} at room temperature.[1] Chemisorption differs from physical adsorption in that gas molecules are distributed in multilayers on the surface, there are only van der Waals interactions between the molecules and the enthalpy of adsorption is quite small (~ -40 kJ mol^{-1}) in the case of physical adsorption. Physical adsorption is thus similar to condensation.[2]

The Langmuir isotherm can be used to relate the concentration of a species A on the surface to the partial pressure of A in the gas phase. It is based on the assumption that the adsorbed molecule/atom is held at defined, localised sites, and that each site can accommodate only one molecule/atom. The energy of adsorption is also a constant over all sites, with no interaction between neighbouring adsorbates.[3] The Langmuir adsorption isotherm can be expressed as:

$$\theta_A = \frac{bP_A}{1 + bP_A} \tag{7.1}$$

where
b is the equilibrium constant for adsorption and $b = k_a/k_d$,
  where
  k_a = velocity constant for adsorption,
  k_d = velocity constant for desorption, and
θ = number of species of A adsorbed.

If dissociation occurs on adsorption as in

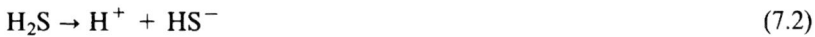

$$H_2S \rightarrow H^+ + HS^- \tag{7.2}$$

then two adjacent sites are required and the equation becomes:

$$\theta_A = \frac{(bP_A)^{1/2}}{1 + (bP_A)^{1/2}} \tag{7.3}$$

The Langmuir isotherm can be expressed graphically as a plot of the amount adsorbed θ against pressure of gas P_A (Figure 7.1).

By expressing θ as V/V_m, where V is the volume of gas adsorbed and V_m is the volume of gas required to give a coverage of one monolayer, the Langmuir isotherm can be rearranged into a linear form:

$$\frac{P}{V} = \frac{P}{V_m} + \frac{1}{bV_m} \tag{7.4}$$

A plot of P/V against P will then be linear with a gradient of $1/V_m$.

After adsorption, the gas phase reactant must undergo chemical reaction at the solid surface and then gaseous products such as water have to be desorbed from the solid and be transported through the pores and back into the gas stream.

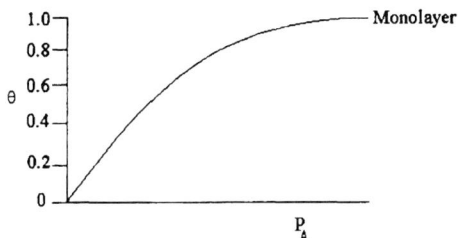

Figure 7.1 *Langmuir adsorption isotherm.*

2 Sorption Kinetics

Mass Transfer and Diffusion

Temperature and concentration profiles are found both inside and outside a pellet in the sorbent bed, and they change continuously until the pellet is completely saturated with adsorbate. As already stated, the sorbate has to be transported to its adsorption site and then reaction products have to be transported back out of the structure. Any of these transport processes could be rate limiting. The overall reaction rate will not be limited by mass transfer if the rates of transfer of gas to the solid surface are much faster than the rate of reaction at the surface. However, if this is not the case then mass transfer effects will be important and the reaction will be diffusion limited. Mass transfer refers to any process in which diffusion plays a role. Diffusion is the spontaneous mixing of atoms or molecules induced by random thermal motion and species diffuse from regions of high concentration to regions of lower concentration. The rate at which diffusion occurs can be expressed in terms of the diffusion coefficient D where D is the rate of diffusion/concentration gradient.[4] The diffusion coefficient is large when the molecules mix rapidly.

The first source of resistance to the path of the sorbate molecule is the boundary film. The boundary film is a thin boundary layer on the surface of the pellet. It is assumed that concentration and temperature gradients between the bulk gas phase and the surface of the pellet are confined to this layer. Diffusion effects through this film are generally negligible, except during the unsteady state conditions that exist when a pellet is first exposed to the gas. However, heat transfer resistance may be important in this boundary layer. The rate of heat generation by transfer of a sorbate from the bulk gas phase through the boundary layer and into the pellet is given by:[5]

$$\text{Rate of heat generation} = k_g \, A \, \Delta C \, \Delta H \qquad (7.5)$$

where
k_g is a mass transfer coefficient/m s^{-1}, *i.e.* it is a measure of the resistance to the
    transfer of material from the external gas phase into the pellet,
A is the external surface area of the solid/m^2,
ΔC is the concentration gradient across the boundary film/mol m^{-3}, and
ΔH is the heat of adsorption for the sorbate/J mol^{-1}.

At equilibrium, heat is lost from the pellet as quickly as it is generated. Then:[4]

$$k_g \, A \, \Delta C \, \Delta H = h \, A \, \Delta T \qquad (7.6)$$

where
h is a heat transfer coefficient/W m^{-2} K^{-1}, and
ΔT is the temperature difference across the boundary film/K.

The most important mass transfer limitation is diffusion through the internal surface of the pellet.[5] There are three possible mechanisms: Knudsen diffusion, bulk diffusion and surface diffusion. The mechanism adopted depends on the pore size compared with the mean free path λ of the sorbate molecules. The mean free path is the average distance gas molecules will travel before colliding. It is given by:[5]

$$\lambda = \frac{1}{\sqrt{2}n\pi\sigma^2} \tag{7.7}$$

where
σ is the collision diameter, the summation of the radii of two gas molecules when they collide (the gas molecules being assumed to be hard spheres), and
n is the number of molecules in unit volume of gas.

Typically, n will be *ca.* 2.5×10^{25} molecules/m^3 at a pressure of one atmosphere at 20 °C and the collision diameter will be *ca.* 0.2 nm. This gives a value for the mean free path of 225 nm. As a rule of thumb, the mean free path should be about 10 times larger than the pore radius at one atmosphere pressure for Knudsen diffusion to occur.[6] Thus, Knudsen diffusion would be associated with mesopores *ca.* 45 nm in diameter. Pores can be classified as macropores ($>$ 50 nm diameter), mesopores (2–50 nm diameter) and micropores ($<$2 nm diameter).[3] Knudsen diffusion is generally restricted to pores $\leq$100 nm in diameter. Under conditions where Knudsen diffusion occurs, a gas molecule will tend to collide with the pore wall before colliding with a second molecule. Collisions between molecules can therefore be ignored and the gas molecule moves down the pore in a random fashion which is interrupted by collisions with the pore walls.

The diffusion of a gas in a pore by Knudsen diffusion is given by:[7]

$$D_K = \frac{2}{3}\sqrt{\frac{8RT}{\pi M}}r_p \tag{7.8}$$

where
r_p is the pore radius,
M is the molecular weight,
D_K is the Knudsen diffusion coefficient in m^2 s^{-1},
R is the gas constant, and
T is the temperature in K.

A process similar to bulk diffusion occurs when the mean free path is small compared with the pore diameter. Bulk diffusion is generally found in pores $\geq$1000 nm. Molecules then collide with other molecules in the gas phase rather than with the pore walls and the diffusion coefficient is independent of the pore radius. This type of diffusion is known as the molecular diffusivity. It is

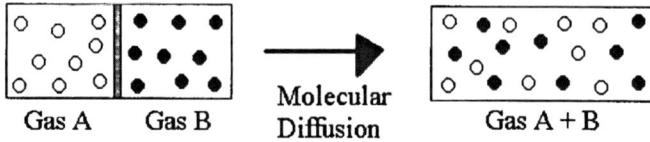

Figure 7.2 *Schematic diagram of molecular diffusion.*

essentially based on Fick's law of diffusion.[8] Fick's law can be explained by considering two gases, A and B, at a fixed temperature and pressure in a given fixed volume, with the two gases separated by an impermeable barrier. If the barrier is then removed, the gases will mix until the mixture is uniform in the given volume (see Figure 7.2).

The mixing process occurs by random motion of the molecules and it is known as molecular diffusion. The rate of diffusion is defined by Fick's law.

Thus, for gas A:

$$M_A = -D_{AB}\frac{dC_A}{dx} \tag{7.9}$$

where

M_A is the number of moles of A per unit area per unit time (known as the molar flux for gas A),

C_A is the number of moles of A per unit volume, *i.e.* the concentration of A,

D_{AB} is the diffusion coefficient for gas A in gas B, and

x is the distance that molecules of type A move during the diffusion process. It is assumed, for simplicity, that the molecules only move in one direction.

Similarly, an expression can be written for gas B:

$$M_B = -D_{BA}\frac{dC_B}{dx} \tag{7.10}$$

where

D_{BA} is the diffusion coefficient for B in A.

Molecular diffusion coefficients can be determined experimentally using Fick's law. However, if experimental values are not available, they can be determined using the Chapman–Enskog equation.[5,7,9]

For diffusion between two gases A and B in a single pore:

$$D_m = \frac{0.0018583T^{3/2}\left(\dfrac{1}{M_A}+\dfrac{1}{M_B}\right)^{1/2}}{P\sigma_{AB}^2\Omega_{AB}} \tag{7.11}$$

where

D_M is the bulk or molecular diffusion coefficient in $cm^2\ s^{-1}$,

M_A and M_B are the molecular weights for gases A and B,

P is the pressure in atm,

T is the temperature in K,

σ_{AB} is the collision diameter in Å,

Ω_{AB} is the dimensionless collision integral. It is a function of kT/α_{AB}, where k is Boltzmann's constant and α_{AB} is the space or void between particles or pellets of sorbent.[7]

The molecular diffusion coefficient for sulfur dioxide in air at 1 atm and 0 °C, for example, is $0.122 \text{ cm}^2 \text{ s}^{-1}$.[9]

Generally, both bulk and Knudsen diffusion can be observed in sorbents since most sorbents contain a range of pore sizes and there is not a sudden change from Knudsen to molecular diffusion at a particular pore size. When both bulk and Knudsen diffusion occur the total diffusivity D_T can be estimated from:

$$\frac{1}{D_T} = \frac{1}{D_M} + \frac{1}{D_K} \tag{7.12}$$

Molecules can also pass into the pellets by surface diffusion.[5] Just as the concentration of gas phase sorbate changes with respect to distance down the pores, there will be a concentration gradient of sorbate on the pore walls provided that adsorption equilibrium is rapidly obtained. Diffusion through this concentration gradient on the pore walls is known as surface diffusion. It is an activated process in which molecules can "hop" between adsorption sites on the pore walls. Surface diffusion becomes important when the surface area of the sorbent or the concentration of sorbate on the pore walls is high, or if the pores are so small that molecular diffusion is negligible.[5,7]

When calculating the total diffusional flow into a pellet allowance must also be made for the fraction of the pellet that is porous and the type of pores present. The summation of the diffusion contributions to flow through all the pores in the porous pellet is thus given by the effective diffusivity D_E:

$$D_E = \frac{D_T \varepsilon}{\tau^2} \tag{7.13}$$

where

D_T is the total diffusion coefficient,

ε is the voidage within the pellet, and

τ is the tortuosity (actual pore length/superficial diffusion path).

τ^2 is referred to as the tortuosity factor and generally has a value between two and six.

In most sorption processes, heat transfer effects within the pellet can be neglected. This is because most adsorptive gases are present in an inert carrier gas and although they have to penetrate the porous structure during adsorption and desorption, the local temperature rise occurring during this process is negligible. It would be necessary to consider heat transfer effects if the reaction was very exothermic.

Breakthrough Curves

The above description describes the interaction between a gaseous adsorbate such as H_2S and a pellet of a solid sorbent. In the clean-up of industrial feedstocks, however, it is important to visualise the interaction of the adsorbate with a bed of sorbent pellets. This is carried out experimentally by looking at the breakthrough curves. The breakthrough curve shows the way in which an adsorbate is distributed when a gas containing a fixed composition of the adsorbate passes into a clean (*i.e.* free of adsorbate) bed until gas emerges in the exit stream. Figure 7.3 shows a diagram of a typical breakthrough rig used for sulfur removal from a feedstock of 2% H_2S in nitrogen.

Slippage of H_2S from a bed of sorbent which is packed in the reactor is detected by the precipitation of lead sulfide in an alkaline lead acetate solution.[10] A precipitate is formed when 2–3 ppm H_2S is present in the exit stream of the sorbent bed. The concentration of adsorbate in both the gas and solid phase at any given point in the bed is a function of time since it results from the movement of the concentration front in the bed. A sharp concentration front is required for an efficient separation. The distribution of an adsorbate along a bed during an adsorption cycle is shown in Figure 7.4.[5] On first introducing the adsorbate into the clean bed of sorbent, the sorbent quickly becomes saturated at the inlet of the bed and is in equilibrium with the adsorbate in the inlet stream. The adsorbate concentration falls off rapidly along the bed and its concentration at the exit is negligible. As the run proceeds, this concentration front passes through the bed due to the progressive satura-tion of the adsorption sites at the entrance of the bed. In Figure 7.4(a), t_1 shows the initial formation of the front, t_2 shows the concentration profile at some intermediate time, and t_3 shows the concentration profile just before break-through, *i.e.* the point at which the adsorbate gas is first detected in the exit

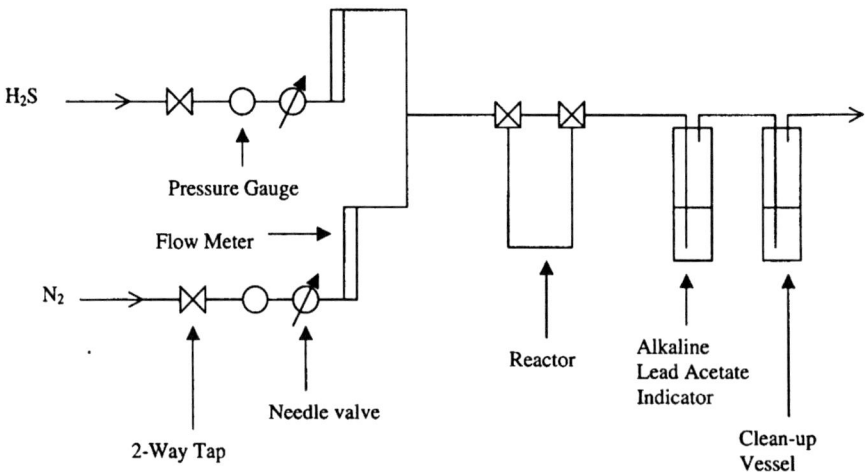

Figure 7.3   *Schematic diagram of a breakthrough rig used for sulfur removal.*

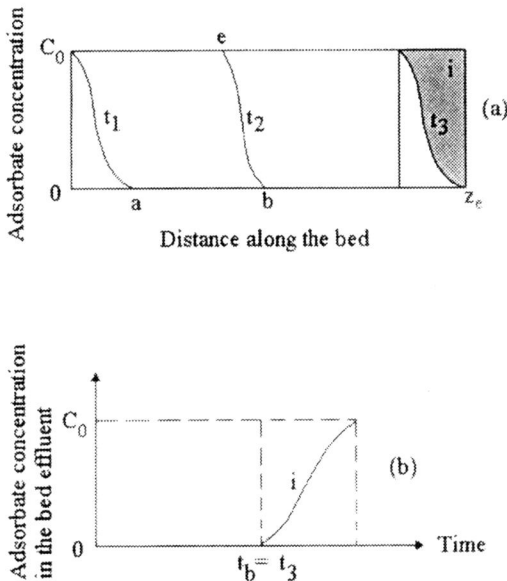

Figure 7.4 *Distribution of adsorbate concentration in the fluid phase through a bed:*
(a) *development and progression of an adsorption wave along the bed,*
(b) *breakthrough curve.*
(Reproduced from ref. 5, p. 781, figure 17.15 by permission of Butterworth
Heinemann, Linacre House, Jordan Hill, Oxford, OX2 8DP. © 1991 J.M.
Coulson, J.F. Richardson, J.R. Backhurst and J.H. Harker)

stream. The run should be stopped just before breakthrough in order to
maximise the efficiency of the sorbent. After breakthrough the concentration
of the adsorbate gas in the effluent stream rises steeply [Figure 7.4(b)].

The shape of the breakthrough curve represents the adsorption kinetics, and
is determined from mass and heat balances on the bed together with the
equilibrium adsorption isotherm. Figure 7.5[5] shows how three different types
of isotherm (plots of gas phase concentration of adsorbate C_g against concentration of sorbate in the adsorbent bed C_s) move through the bed of sorbent.

In case (a), the isotherm is concave and it is termed favourable as the
adsorption zone becomes narrower as it passes through the bed. This is because
there is a distribution in the speed with which points of high and low
concentration move through the bed, the high concentration zones moving
faster but being unable to overtake the low concentration zones further down
the bed. In case (b), the isotherm is linear so there is no concentration variation
with respect to distance in the bed and the adsorption zone passes through the
bed unchanged. In case (c), the isotherm is convex. The adsorption zone
increases in length as it moves through the bed as the lower concentration zone
now moves the fastest and the isotherm is said to be unfavourable. Even in
favourable cases, some 'spreading' of the adsorption zone will be observed due
to resistance to mass transfer and longitudinal dispersion. Longitudinal dispersion arises from the dispersion of the sorbate axially in the bed and it is the

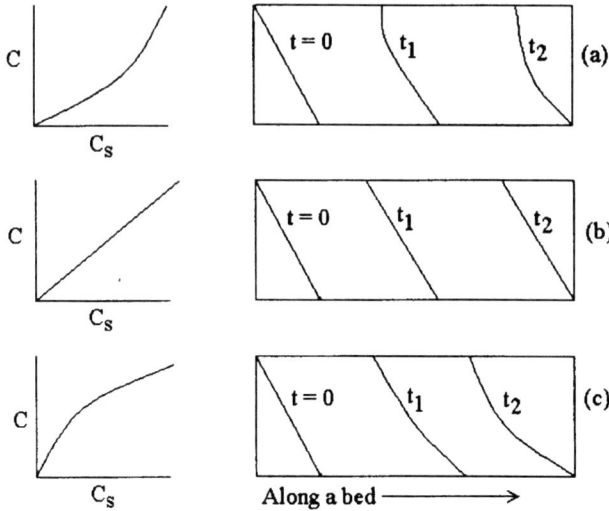

Figure 7.5 *Effect of the shape of the isotherm on the development of an adsorption wave through a bed with the initial distribution of adsorbate shown at t = 0.*
(Reproduced from ref. 5, p. 785, figure 17.18 by permission of Butterworth Heinemann, Linacre House, Jordan Hill, Oxford, OX2 8DP. © 1991 J.M. Coulson, J.F. Richardson, J.R. Backhurst and J.H. Harker)

product of the dispersion coefficient and the concentration gradient in the bed. The dispersion coefficient can be measured by injecting a pulse of a tracer into the bed and measuring its change in shape between two predetermined points in the bed.[5]

The equilibrium isotherm can be incorporated in the kinetic model that is used to model the transport of an adsorbate through the sorbent bed. A mass balance is carried out for the adsorbate in the gas stream flowing through a small cross sectional area through the bed δz. It is assumed that the rate of loss of sorbate by adsorption from the gas phase is equal to the rate of gain in the adsorbed phase. It is also assumed that the adsorbate concentration in the inlet stream is low, which then means that the gas flow can be regarded as constant along the bed. Applying these assumptions it can be shown that:[5,7]

$$\left(\frac{\delta z}{\delta t}\right)_c = \frac{u}{1 + \left[\frac{(1-\alpha)}{\alpha}\frac{dC_g}{dC_s}\right]} \tag{7.14}$$

where
δz refers to an increment of the bed of sorbent analysed with respect to time δt,
$(\delta z/\delta t)_c$ is the velocity of the concentration front in the sorbent bed,
u is the flow rate of the gas between the pellets in the fixed bed in ms^{-1},
dC_g/dC_s is the slope of the adsorption isotherm of gas phase concentration of adsorbate against adsorbate concentration in the bed, and
α is the voidage between the pellets in the bed.

The breakthrough time, *i.e.* the time taken for the adsorbate to emerge from the sorbent bed, can be determined easily for the simplest case in which the isotherm is either favourable or linear and in which there is no axial dispersion.[7] It is assumed that the sorbate is admitted to a clean bed as a constant concentration plug at time zero.

In practice, the adsorption zone will not remain as a plug but will disperse as it moves through the bed. The breakthrough time can then be determined by integrating equation 7.14 at constant *C* (concentration of adsorbate in the gas stream).[5,7]

Thus, integrating equation 7.14 and rearranging, the breakthrough time *t* is given by:

$$t = \frac{z - z_0}{u}\left[1 + \frac{(1 - \alpha)}{\alpha}\frac{dC_g}{dC_s}\right] \tag{7.15}$$

where
z_0 is the position in the sorbent bed where the adsorbate is initially found (thus for a bed that is initially free of adsorbate, z_0 is zero for all values of *C*),
z is the length of bed in m, and
t is the breakthrough time in s.

It is necessary to modify the above equation to take into account longitudinal dispersion effects which are likely to be significant at the low flow rates necessary for the adsorption process to approximate to equilibrium adsorption. Under these conditions it has been shown that:[5]

$$\frac{C}{C_0} = \frac{1}{2}\left\{1 + \mathrm{erf}\left[\left(\frac{uz}{4D_L}\right)^{1/2}\frac{(t - t_{min})}{(tt_{min})^{1/2}}\right]\right\} \tag{7.16}$$

where
C_0 is the constant concentration of adsorbate in the gas entering the bed,
D_L is the longitudinal diffusivity in $m^2\ s^{-1}$,
t_{min} is the minimum time required to saturate a bed of unit cross-section and length *z* for a given flow rate *u*.

Also:

$$t_{min} = \left[1 + \left(\frac{1 - \alpha}{\alpha}\right)\frac{C_{s\infty}}{C_0}\right]\frac{z}{u} \tag{7.17}$$

where
$C_{s\infty}$ is the concentration of adsorbed phase in equilibrium with C_0,
t is the actual breakthrough time, and
erf is the error function.

$$\mathrm{erf}\ x = \frac{2}{\sqrt{\pi}}\int_0^x e^{-x^2} \tag{7.18}$$

Values for erf and its derivative are given in standard tables; see, for example, ref. 8.

The equilibrium case cannot be used for breakthrough curves where the flow rate of the adsorbate is considerable and/or the sorbent pellets are greater than 250–420 μm in size.[7] Larger pellets are generally used in industrial reactors to minimise the pressure drop across the bed, and it is then necessary to include mass transfer effects in expressions to model breakthrough curves. Mass transfer resistance in the pellets has been determined using the Rosen model[11] and the Thomas model.[12] In the Rosen model, the sorbate pellet is considered to be homogeneous and the rate of mass transfer in a clean bed of sorbent is determined from a combination of external film and internal pore diffusion. It is also assumed that the isotherm is linear and isothermal, there is no axial dispersion and both the flow and effective diffusivity are constant. In the Thomas model, it is assumed that the rate is controlled by the surface reaction and that diffusion in both the film and pores takes place instantaneously. A Langmuir adsorption model is assumed to apply in this case. Both of these models can predict breakthrough behaviour to within 1% of that observed experimentally.[7] A quantitative treatment of the Rosen model is given in ref. 7 and ref. 5 and the Thomas model is detailed in ref. 7.

3 References

1 T. Baird, K.C. Campbell, P.J. Holliman, R.W. Hoyle, D. Stirling and B.P. Williams, *J. Chem. Soc., Faraday Trans.*, 1995, **91**(18), 3219.
2 M.J. Pilling and P.W. Seakins, 'Reaction Kinetics', Oxford University Press, Oxford, 1995, chapter 7, p. 175.
3 G.C. Bond, 'Heterogeneous Catalysis', Oxford University Press, Oxford, 1987, chapter 2, p. 12.
4 P. Atkins, 'The Elements of Physical Chemistry', 2nd edn, W.H. Freeman and Company, New York, 1997, chapter 10, p. 383.
5 J.M. Coulson, J.F. Richardson, J.R. Backhurst and J.H. Harker, Chemical Engineering vol. 2, 'Particle Technology and Separation Processes', 4th edn, Butterworth Heinemann, Oxford, 1998, chapter 17, and references therein.
6 A. Wheeler, *Adv. Catalysis*, 1951, **3**, 249.
7 R.T. Yang, Chemical Engineering vol. 1, 'Gas Separation by Adsorption Processes', Imperial College Press, London, 1997, chapters 4 and 5.
8 J.M. Coulson, J.F. Richardson, J.R. Backhurst and J.H. Harker, Chemical Engineering vol. 1, 'Fluid Flow, Heat Transfer and Mass Transfer', 5th edn, Butterworth Heinemann, Oxford, 1998.
9 J.R. Anderson and K.C. Pratt, 'Introduction to Characterisation and Testing of Catalysts', Academic Press, Sydney, 1985.
10 R.W. Hoyle, 'Low Temperature Gas Desulfurisation using Mixed Cobalt-Zinc Oxides', PhD Thesis, Glasgow, 1995.
11 J.B. Rosen, *J. Chem. Phys.*, 1952, **20**, 387.
12 H.C. Thomas, *J. Am. Chem. Soc.*, 1944, **66**, 1664.

Subject Index

CPSIA information can be obtained at www.ICGtesting.com
Printed in the USA
LVOW031634311011

252873LV00008B/88/A

9 780854 045419